Praise for *Saving Energy, Growing Jobs*

"Energy efficiency, climate protection, and clean air all make sense and make money, because saving energy costs less than buying it. That's why smart companies boost earnings by using six percent less energy each year per unit of production—laying off idle kilowatt-hours and gallons so they can retain their employees. Now a leading creator of successful public policies that support this business logic distills his decades of experience into a compelling summary. David Goldstein shows how applying these insights more widely can create prosperity, jobs, and real security, and how sloppy logic continues to underpin false dogmas holding the opposite."
> **Amory B. Lovins,** Cofounder and CEO,
> Rocky Mountain Institute

"David Goldstein has written an enormously valuable and challenging book—challenging to business and also challenging to environmentalists. Everyone concerned about the future of the economy and the future of the environment should read this book. Better than any other, it explains how to move beyond the myths of the past."
> **James Gustave Speth,** Dean, School of Forestry and
> Environmental Studies, Yale University

"A pioneer of energy efficiency policy, David Goldstein documents how energy efficient appliances, far from bankrupting their manufacturers, have led to more profitable companies. Goldstein, with concrete examples, shows that the goals of business leaders and environmentalists can be served simultaneously by well-crafted policies, and that we will all reap the benefits."
> **Robert Socolow,** Professor of Mechanical and
> Aerospace Engineering, Princeton University

Saving Energy Growing Jobs

Saving Energy Growing Jobs

How Environmental Protection Promotes
Economic Growth, Profitability,
Innovation, and Competition

David B. Goldstein

Bay Tree Publishing
Berkeley, California

© 2007 David B. Goldstein

Cover design by SueRossi.com

Library of Congress Cataloging-in-Publication Data
Goldstein, David B., Ph.D.
   Saving energy, growing jobs: how environmental protection promotes economic growth, profitability, innovation, and competition / by David B. Goldstein.
       p. cm.
   Includes bibliographical references and index.
   ISBN-13: 978-0-9720021-6-5 (alk. paper)
   ISBN-10: 0-9720021-6-2 (alk. paper)
   1. Energy policy--United States. 2. Energy con-servation--Environmental aspects--United States.
3. Energy conservation--Economic aspects--United States. 4. Environmental protection--United States. I. Title.

   HD9502.U53G65 2007
   333.79'160973--dc22
                                        2006025664

This book is dedicated to Julia Vetromile, my wife, who (with an appropriate mix of subtlety and assertiveness) made sure that it got written.

಄

Contents

Foreword

I have long held a vision of politics and public life as positive and constructive endeavors and believe in reaching out to bridge consensus to make the system work. But increasingly, and regrettably, debates on the direction of the American economy and our environment have become more polarized.

Solutions have taken a back seat to sound bites, when it is the merits of an argument and the worthiness of a cause that should determine the course of events in government. As a result, the nation and the world at large is missing countless opportunities both to clean up our environment and spur economic growth.

Saving Energy, Growing Jobs offers a compelling way forward, and demonstrates how environmental protection is not merely compatible with economic growth, but can also create business opportunities. A healthier environment can and must come from collaborative relationships among government, private industry, and the environmental movement. David Goldstein portrays how those relationships can be developed and how they can foster products that can constitute a legacy for the common good, leaving a minimal "footprint" on the world we leave to future generations.

The key to helping business thrive and nurturing a healthy environment is innovation, which has been the bedrock of this nation's growth since its inception.

This book argues that the government must actively promote innovation. The United States has a long-standing history of creativity and ingenuity, but when it comes to energy and the environment, we have not relinquished our dependency on foreign oil as the dominant energy source, nor have we been able to halt damaging changes in the earth's climate. For the energy security of our nation, we must begin a new chapter for energy use in these threshold years of the twenty-first century. We require fresh sources for generating energy and new means for saving it, but we cannot achieve that goal without active governmental policies.

Innovation, as David Goldstein posits, does not appear automatically just because we have a market economy. Markets can fail in both simple and complex ways, and these failures stifle innovation while allowing unnecessary pollution, waste, and environmental damage to continue.

Businesses are especially impacted by our nation's inefficient use of energy. Generally, the more energy a business requires, the less profitable the business. In order to grow our economy, we must not burden it with crippling energy costs. Part of the solution to avoiding that calamitous route lies in reducing energy use while increasing productivity.

Saving Energy, Growing Jobs describes how companies often do not identify and correct wasteful uses of energy and how good government can minimize or even eliminate failures that lead to a loss of competitiveness. This book suggests, contrary to accepted opinion, that better communication between business leaders and environmentalists can contribute to job creation and promising opportunities for small and large businesses while furthering environmental goals that will benefit us all.

Unfortunately, rigid theories and calcified preconceptions impede progress. David Goldstein conveys a more practical approach, encouraging business leaders, government officials, and environmentalists to adopt open approaches to issues without the hindrance of partisan bias. Focusing on real-world experience, the author leads us to conclude just how environmental protection can simultaneously enhance market forces and promote economic growth.

A particularly vital area of environmental and economic concern is the increasingly apparent need to limit human-induced climate change. Ongoing scientific peer-reviewed research overwhelmingly demonstrates that climate change is one of the watershed issues of the twenty-first century.

Mr. Goldstein asserts that controlling climate change can be achieved at no net cost to the U.S. or world economy and that limiting climate change can be part of an international economic development policy—if it is done correctly.

On Wednesday, February 16, 2005, the Kyoto Protocol on climate change officially entered into force, with 141 countries and regional economic integration organizations depositing instruments of ratification, accessions, approvals, or acceptances with the United Nations. Requiring mandatory cuts in greenhouse gas emissions by 35 participating developed countries starting in 2008, the treaty represents the beginning of true international action on climate change.

With the aid of this book, we are learning that securing international cooperation is indispensable, as we cannot get to the heart of this global problem without the world's major economies communicating at the table. The United States and Australia have not ratified the Protocol, and major rapidly industrializing countries, such as China, India and Brazil, while signatories to Kyoto, are not required to make cuts for carbon dioxide. The causes of climate change are global and the atmosphere knows no political boundaries, so the challenge we face for the future of the planet must be met with all the countries of the world working together.

This is the main reason I agreed to co-chair the International Climate Change Taskforce over the past two years. Our 2005 report, Meeting the Climate Challenge, recommends ways to involve the world's largest economies in the effort, including the U.S. and major developing nations, focusing on creating new agreements to achieve the deployment of clean energy technologies, and a new global policy framework that is both inclusive and fair.

Our country, and much of the developing world, must recognize that continuous improvement in energy efficiency and new non-carbon-emitting sources of energy can and must be developed and launched in the marketplace. Moreover, competition will not only improve technologies, but will also make them more available to more nations.

Saving Energy, Growing Jobs challenges us to go beyond the myths of what businesses or environmentalists want—to discard the allure of the zero-sum game. Unquestionably, we all desire economic growth. We all want to protect our environment and minimize climate change. This book reminds us of the dire pitfalls of either-or propositions and instead suggests how we can approach these issues more effectively. Embracing a smarter sensibility toward policies, incentives, taxes, regulations, and building codes is a critical step forward. And the good news is, we already know how to achieve this end.

Consider California, New York, and Massachusetts, where these economies use dramatically less energy per capita than the United States as a whole. Business prospers while the environment improves—to borrow an iconic phrase from Humphrey Bogart ... that could be the beginning of a beautiful friendship. And this book moves us closer to fulfilling that aspiration by engendering a positive outlook. There should be no mistake, global warming as well as widely fluctuating crude oil prices are real, and serious. The world's excessive misuse of energy adversely impacts billions of people worldwide.

Intelligent choices are at our disposal—that is the theme of this book written with a fresh, accessible pragmatism. *Saving Energy, Growing Jobs* lays out pathways to achieve economic growth and create positive economic impacts. Best of all, it offers encouragement to our young scientists, engineers, environmentalists, and business people, suggesting that together we can generate economic growth and a cleaner environment through innovative, practical solutions.

—Senator Olympia J. Snowe

Preface

In this book I draw on my personal experience working as an advocate of energy efficiency policies to raise a number of issues that I believe have been almost totally overlooked, both in the policy dialogues in the United States (and elsewhere), and also in the academic discussions of energy, environment, and the economy.

My purpose is to suggest new approaches to political consensus on the environment—approaches that can make business more profitable, that will produce more and better jobs, and that will accelerate economic innovation and growth. These approaches rely on market-based governmental policies to protect the environment, primarily through new technology.

Economics is unlike other sciences in that its explicit goal is to guide public policy. In this sense I have written this book within the spirit of mainstream economics. The book is also within the mainstream of economic thought in that I recognize the important role of market forces in guiding economic and environmental policy.

What is unique in the book is my observation that to rely on and to enhance market forces is more complex than previously believed. Economic theory does not state that an increase in the role of market forces necessarily accompanies a decrease in the role of government, or even in the role of government regulations. Instead, in many cases an active and continuing governmental role is necessary to secure the effective operation of markets.

A careful examination of economic theory and economic data demonstrates how environmental protection has generally promoted economic growth even when that outcome was unintended, and suggests how and why this growth opportunity will be even larger if proponents of growth and advocates for the environment can work together.

The stakes are high for both the competitive position of the United States in the global economy and for world economic prosperity and environmental health. An environmentally sound energy policy could boost productivity by over tens of trillions of dollars worldwide and create greenhouse gas emissions reductions of well over 75 percent compared to common forecasts. Environmentally destructive policies can compromise growth, as California's experiment with electricity "deregulation" in the 1990s illustrates.

I draw on a wide variety of academic disciplines and practical experiences in this book. This presents a problem for both the author and the reader. The ideal approach to this project would have been to write two books: one that addresses the issues from an academic and rigorously scientific perspective, and another that takes a more practical look at the real-world consequences of the more rigorous and less-accessible volume.

I take only the latter approach, although in parts of it I try to find a middle ground. I do this for several reasons. First, the academic literature on the subjects I cover is extremely thin, particularly in the attempt to make connections between fields. That is to say, limited written information exists on the subject, and most of what is written is scattered: it consists of three-paragraph discussions in the middle of 400-page books. To take a thorough and rigorous approach would require years of work and active collaboration between experts in a variety of departments and fields. One of my purposes in this book is to inspire such new research.

However, the consequences of my hypothesis, if it is correct, are too important to wait for all the completed research to come in. And unless this hypothesis is raised and debated in advance of the results, the appropriate work will not be performed in any event.

Science is not a random search for facts. It is the attempt to support or disprove well-framed hypotheses.

Before a question receives an answer, we must first carefully frame it. In many cases, how the question is framed determines the types of answers derived. For example, if astronomers ask the question, as they did in medieval times, "How does one predict the motion of the sun and the moon and the stars around the earth?" we can produce the fifteenth-century answer that assumes the stars, moon, and sun all rotate in fixed cycles around the earth, modified by epicycles and second- and third-generation epicycles.

We will have answered the question adequately in terms of predicting the positions of the celestial bodies. Yet we will have totally failed to arrive at the truth. Furthermore such a framing of the question fails even to suggest the necessary research to arrive at the right answer—that the earth and the planets revolve around the sun and only the moon revolves around the earth.

Instead if the question is reframed as, "How does one explain the apparent movement of the sun and moon and stars with a few simple and universal laws?" we get the correct scientific answer about the orbits of planets and moons as Newton's laws of gravity and motion explain.

I attempt in this book to reframe key questions about how real economies function and how environmental policies affect these economies. How I reframe the questions will lead different readers to different conclusions.

The general-interest reader, or the business reader, can find in this book some reasons why greater dialogue and joint policy recommendations between business interests and people who want to enhance economic growth, on one hand, with environmental advocates, on the other hand, should be fruitful. These reasons require we delve into a number of technical issues in order both to understand where the common interests lie and why these problems have remained unsolved for decades. The technical issues may be of great interest to some readers, and less to others. They are included in the detail they are because the main arguments will be unclear without this detailed evidence.

So when confronted with details that concern the California energy crisis of 2000 or about the implicit assumptions that exist in economic theory, the reader is asked to recognize that these discussions are provided not only for their inherent interest but also because they illustrate points that are pivotal to the policy discussion. Such discussions are necessary to address the question of why so many pro-business interests automatically oppose environmental-protection proposals. They illuminate the issue of how representatives of particularly narrow economic interests are able to get political traction for policies that have so few real beneficiaries.

The academic reader should recognize this book as a presentation of one main hypothesis, and several subsidiary hypotheses, that I have framed rigorously enough to allow serious research regarding them, along with some evidence to make them plausible. The academic reader will be disappointed at the relative lack of references compared to similar books. This lack is only partly my fault; many of the most important issues have almost no relevant literature. For other issues, the literature is scattered, and as I discussed above, a comprehensive list of references would cite hundreds or even thousands of articles and books, with main points that are tangential or irrelevant to my hypotheses, and only small parts of which are important to these hypotheses.

I find the weakness in literature surprising, but if the questions are framed the way I have done, such is the case. Indeed throughout much of the book I attempt to question why this is so.

Extensive literature on subjects with similar titles (or even abstracts) does exist; however, almost all of this literature that I have been able to review addresses issues that are not germane to the ones I discuss in this book, or offers similar-sounding issues that are based on assumptions—often implicit yet equally constricting as if more direct—that render the results irrelevant to shed light on the issues I raise in the book.

So a literature review would be exceptionally lengthy and tedious to read; it would consist of a long list of seemingly related books and publications, followed by explanations of why they truly fail to deal with the issues at hand.

I also explore the questions of how markets for environmentally related production and consumption work in real life and how they are described in theory. I identify deep disconnects between what is represented as theory versus what is reality. As I tried to explore the reasons for this disconnect and to explain what was observed in terms of basic economic principles, I found many deep failures of communication that led to what appear to be needless controversies about the environment, often founded on deeply felt myths that guide the perceptions of both business and environmentalists.

One of the major themes contained in this book is a critique of the misuse of economic theory to direct policy debates. Economics is intended to be used to guide action as well as to explain observations. I take the same approach: my goal is to explore new directions in economic policy that can use environmental-protection objectives as a way to enhance economic growth and development.

Most everyone wants to create economic growth and an expansion of job opportunities in his or her own region, nationally, and globally. Most everyone wants a cleaner and more healthful environment. Most political leaders, and both political parties in the United States, claim we can have both. Yet few political leaders advance actual policies that will get us there, and in many cases the broad, sweeping claims of pro-growth and pro-environmental attitudes of both American parties are undercut by the actual proposals for laws or regulations that fail to do a good job at meeting either objective.

In this book I attempt to address such a failure: I point out realistic opportunities and policies that ought to be enacted with strong consensus. I am optimistic about the prospects for this change, based on five years' effort to lobby on a bipartisan proposal for tax incentives for energy efficiency in buildings. I discovered, once I had the opportunity to explain the proposal, that the arguments for a well thought-out bill were equally attractive to the far right and the far left. Sometimes one political extreme had valid arguments that influenced the shape of the legislation that also appealed to the opposite extreme. (The bulk of the bill was incorporated into the Energy Policy Act of 2005.)

I wrote this book based on my experience in environmental advocacy, and indirectly based on my wife's experiences as an engineer and a manager in both industry and the consulting business. Naturally the book reflects some of the imbalances or even biases to which this experience would lead. However, it is neither written on behalf of environmentalists nor an individual environmental organization or program. No environmental-protection advocacy organization has reviewed the drafts, nor has any business organization or private company.

This book reflects my belief as an individual who tries to be objective and to listen to and evaluate all sides of an issue. It is my attempt to see beyond the disagreements that seem to characterize almost all of the environmental discussions in the United States, and look for the vast potential for common ground. To some extent, this broader vision is increasingly visible in the relationship of business to environmental issues in other countries, but to a larger extent this approach is likely to be a new one everywhere.

The world faces immense challenges over how to raise the standard of living of some four billion people whose economic prospects are now dismal. We must do this in the context of levels of pollution and destruction of natural ecosystems that is already unsustainably high and is generally expected to worsen. If the world endeavors simultaneously to solve each problem as though it were a competing demand on limited resources, we will likely fail at both objectives. However, if we recognize that we can solve both problems with basically the same approach, we have grounds for optimism.

Acknowledgements

In my efforts to write this book I have drawn on a wide variety of sources of support, intellectual content, discussion, and experience and I almost hesitate to mention any names of those whose contributions were essential to its completion because I know that I may fail to recognize all the people and institutions to whom I am indebted.

First, the actual drafting of this book follows a graduate seminar I taught at the University of California's Energy and Resources Group in Spring 2004. The Physics Department at the University of California at Berkeley supported the transcription of the book, which was performed with great speed and accuracy, as well as valuable editorial suggestions, by Evelyn Arevalo.

This seminar could not have occurred without the support of Dan Kammen, Chris McKee, Mark Richards, Virginia Rapp, and Zack Powell, all of the University of California at Berkeley. Dan also reviewed the draft manuscript and gave me valued guidance.

The students in the Energy and Resources seminar, including those formally enrolled and those who attended classes unofficially, contributed greatly to the development of ideas both cited and not cited in the text. They are Lorraine Lundquist, Beth Zotter, Bill Roller, and Chris Jones; also Hung-Chung Fang, Robert Crockett, Tyler Dillavou, Harish Agarwal, and Barbara Haya.

Many of the initial ideas contained in the book were developed in partnership with Alan Miller, who initially worked as my attorney and colleague at the Natural Resources Defense Council (NRDC), and later as my co-officer of the Institute for Market Transformation. Alan and I spent appreciable time following our discussions of the issues concerning how to advocate appliance efficiency standards—which is what we were supposed to do—trying to work more deeply to understand the misuses of economic theory, why they were so pervasive, and what it would take to confront them.

Alan introduced me to the work of Steve DeCanio, whom I spoke with much later but whose ideas contributed strongly to this book. Much of the technical discussion on economic fundamentalism is based on the insights I found in his book *Economic Models of Climate Change: A Critique* and in our telephone conversations. And Alan also reviewed the preliminary draft of this book and gave me advice on how to move it foreward.

Ashok Gadgil, now at the Lawrence Berkeley National Laboratory, provided additional encouragement and original ideas.

The basic experience upon which this book draws began with my work with Art Rosenfeld, whose career has taken him from the University of California and the Lawrence Berkeley National Laboratory, where I first made his acquaintance, to the U.S. Department of Energy (DOE) and the California Energy Commission (CEC). Art taught me more than I can ever even realize. How he thought and worked affected my work so profoundly that I often failed to recognize his influence until many years later. Art's advisor at the CEC, John Wilson, also shaped how I perceived the role of regulation and incentives in a market economy from the onset of my professional relationship with him around 1980.

My introduction to the broader issues of economic development and efficiency was based on a course at U.C. Berkeley's Department of City and Regional Planning that Richard L. Meier taught. I additionally benefitted from the overall approach and thought processes to which John Holdren and John Harte of the Energy and Resources Group and Rob Socolow of Princeton and Marc Ross of the University of Michigan introduced me.

I also want to thank my NRDC colleagues Ralph Cavanagh, David Edelson, Peter Miller, Ellie Goodwin, and later Sheryl

Carter, Noah Horowitz, Kit Kennedy, Ashok Gupta, Rob Watson, Dale Bryk, Dan Lashof, Bob Fisher, Bob Epstein, Evelyn Arevalo, Shari Walker, Devra Bachrach Wang, and John Walke, for the continuing dialogue on economic issues that eventually led to the ideas expressed in the book.

I am also grateful to my colleagues in industry, government, and in other public interest organizations who helped with the concepts: Mike Thompson, Chuck Imbrecht, and Jeff Johnson (who all passed away far before their time), Gary Fernstrom, Harvey Sachs, John Hoffman, Kathleen Hogan, Michael L'Ecuyer, John Fox, Cathy Zoi, Liz Klumpp, John Millhone, Tom Graff, Gene Rodriguez, Jonathan Blees, Bill Pennington, Tom Eckman, Margie Gardner, Tom Foley, Bruce Wilcox, Lee Schipper, Gerry Groff, Alan Meier, Yuri Matrosov, Edward Volkov, Katherine Buckley, Charles Eley, Doug Mahone, Mike Gabel, Philip Fairey, Chuck Samuels, Joe McGuire, Joe Mattingly, Jack Langmead, Alex Shively, Warren Weinstein, Rachel Miller, Louise Dunlap, Joe Browder, Ginny Worrest, Elizabeth Paris, Jeff Duncan, Tim Charters, Kathleen Shields, Greg Parks, and Manny Rossman.

A number of other colleagues also contributed to my understanding of the issues of economy and environment. I particularly want to acknowledge the leadership of Skip Laitner and Curtis Moore, and especially the assistance of David Driesen (who kindly reviewed the draft of the entire manuscript and offered valued comments). The issue of the California energy crisis has seldom been described from start to finish; I appreciate the advice of Andy Van Horn, who edited the initial Chapter 5 draft.

Many of my friends and colleagues helped me successfully launch a book-publishing project. I want to thank Bill Roller in particular for his audit of my U.C. Berkeley class and his encouragement to write this book. I also thank Bill because he read my earlier drafts and guided me through the book-publishing process. Gus Speth and Joe Romm also provided valuable advice on how to write one's first book. I also wish to acknowledge my publisher, David Cole, who put far more effort into this project than any of the publishers my friends have experienced, and my editor, Alan Rinzler, who gave me lots of good suggestions that I wanted to ignore yet am grateful I refrained.

Some of the most important experiences that gave me insights into how politics interacts with economics came from my work with Congressional offices on legislation to provide tax incentives for energy-efficiency buildings. So I would like to thank Senators Olympia Snowe and Dianne Feinstein, along with former Senator Bob Smith, and Representatives Randy ("Duke") Cunningham and Ed Markey, and those members of their staffs whom I neglected to name above.

Readers might initially suspect that the story I present is merely about energy. Yet several excellent books on energy efficiency already exist, of which perhaps the most recent is written by my colleague, Howard Geller, former executive director of the American Council for an Energy Efficient Economy (ACEEE). Many of the experiences that led to this book were collaborative endeavors with Howard and his successor, Steve Nadel.

The actual writing of this book was only made possible with the determination of my wife, Julia Vetromile, who inspired me to write it this year and not next. The text reflects the amount of influence she has had on the content through sharing her engineering experiences and through developing and refining the material over coffee and on long walks. From my early childhood, in my family—and my wife's—children were and are always part of continuing policy and political discussions, and so I also want to thank Elianna Goldstein and Abraham Goldstein for the ideas they contributed, both directly and indirectly, and for how they stopped me whenever they perceived something as nonsense. I also offer thanks to my parents, Laurence and Gloria Goldstein, and my in-laws, John and Mary Jane Vetromile, for conversations that taught me to understand both conservative and liberal perspectives on policy issues, and how to respond to different positions in a reasoned way.

This book also is the direct result of the MacArthur Fellowship program. The trustees of the MacArthur Foundation established this program to encourage creativity and innovation, and I was fortunate enough to receive one of their fellowship grants in 2002. The grant was intended to promote creativity. For some recipients this may be straightforward, but for me I kept thinking that they were asking me to stand in a corner and be creative.

However, the fellowship did encourage me to step back from the direct experiences of my almost thirty years of environmental advocacy and think bigger—and farther outside the box. Many people asked me, after I received the MacArthur fellowship, whether I would take some time off. I responded truthfully that I was already doing what I most desired with my career.

Yet at the time I also recognized my failure to take advantage of the opportunity to apply my experience to the larger economic/political issues that I had become awakened to in the context of my work: the issues of how a market economy actually works—as opposed to how the economic fundamentalists say it works; of how regulations and incentives fit into the context of markets; and of how trade associations affect the market.

I discovered that the issues mentioned above had not been addressed fully—or even at all—in the academic literature, much less the policy dialogue. As a result, business leaders and political leaders lacked exposure to a balanced discussion of environmental policy issues. This book is my response to this lack.

Introduction

No one supports pollution as public policy. No matter what one's political philosophy or business goals, clean air, safe drinking water, foods free from poisons, and the protection of unspoiled environments are seemingly universal goals.

Then why is environmental policy so controversial? The only plausible policy reasons to oppose environmental protection, and indeed, the primary arguments anti-environmental advocates raise on virtually all of the major environmental controversies of the last several decades, are the beliefs that a clean environment...

1. Requires unacceptable compromises in our economic well-being, or
2. Places unreasonable restrictions on human freedoms.

This book demonstrates how such concerns are ill founded. The arguments and evidence I offer instead support how and why well-designed policies for environmental protection will enhance economic development, create greater employment, and provide more democratic ground rules for the economy.

The Case for Environmental Protection

When the environmental movement in America began to take strong advocacy positions, after 1970, opponents of environmental protection argued that there was a fundamental tradeoff between economic growth and prosperity and the environment: the more we protected the environment, the more sacrifices we would have to make economically.

Economic theory allegedly supports such standard wisdom—that environmental quality competes with economic growth.

Most of the anti-environmental arguments of the past, and many of the critical arguments that continue into the present, are theory-based economic arguments about how a particular environmental policy will lead to any or all of the following conclusions:

1. Hurt business. This is the most frequently heard argument against environmental protection. Environmental laws are claimed to restrict business's choices, raising costs and cutting profits and growth.
2. Hurt consumers—particularly poor people—or increase poverty. If environmental laws increase the costs of basic necessities, this will arguably affect the poor disproportionately. (The latter argument is perhaps the least convincing because low-income advocates and consumer groups rarely oppose environmental laws, and often support them.)
3. Compromise freedom or limit property rights. Opponents of environmental protection sometimes argue that environmental policies limit market choice or restrict property rights.

However, a closer examination of economic theory actually supports the opposite of such conclusions: that environmental protection policies can enhance economic growth, help consumers (and specifically the poor), and reduce restrictions on individual and corporate freedom.

Both theory and practical experience illustrate how environmental policies:

1. Reduce costs to businesses and consumers
2. Increase productivity by encouraging new technologies
3. Overcome systematic failures of real markets to maximize profits and in particular to promote innovation
4. Enhance competition by breaking up nonmarket relationships between a few industry-leading companies (that frequently have large and stable market shares for their products)

Businesses and their political allies consistently have argued that environmental regulations and incentives compromise profitability and growth. However, for the last few years both political parties officially have been supportive of environmental protection and maintain that we can have both economic growth and environmental protection at the same time. (Visit each of their websites.) But these arguments are based mainly on hopes, without reference to facts and experiences.

So what are the facts with respect to the impact of environmental policy on economic growth? Considering the great significance of this question, I will show how the economic benefits of only a few of the many environmental policies are in the trillions of dollars—surprisingly little solid research on this subject can be found. The few studies that have been done generally support the view taken here, but they have mostly focused on a narrower and more difficult hypothesis: that all environmental regulation promotes economic growth. Documentation of the benefits of environmental protection, which includes regulation but also involves other policies that protect the environment, will necessitate further research and study. This book sets forth some of the evidence that is available now, and suggests some directions for the research effort.

The Relationship Between Environmental Protection and Economic Growth

Increased evidence, based on almost forty years of experience in strong environmental protection policy in the United States and globally, supports the assertion that carefully designed environmental protection policy affirmatively promotes economic growth. I provide specific examples of environmental regulations and other policies that have had unexpectedly large economic benefits—independent of the direct environmental benefits. Environmental regulation can promote innovation—the engine of economic growth for the twenty-first century—and enhance competition.

I suggest that well-designed environmental policies, including regulations as well as incentives, can both spur innovation and overcome failures of the marketplace. They break apart cozy, anti-competitive relationships between large corporations, and promote broader and deeper competition, and the more effective use of market forces. They can also enhance personal freedom and democracy by eliminating or revising private-sector regulations that limit economic choice.

Failures of the market are more widespread and systematic than is generally understood, and the influence of intrusive private-sector regulations, often written by established companies, is larger than many people realize, so to free the markets and promote competition through environmental policies could be surprisingly effective.

Additionally, more competitive market structures that result from environmental protection policies can promote economic growth by encouraging innovative thought, product development, and changes in industrial processes to make them more productive, more profitable, and cleaner. Improving productive processes yields more jobs; and the activity of improving products and processes produces higher-paying jobs. More transparent market rules and regulations protect our freedoms by reducing the ability of a limited number of companies with strong economic power to limit the choices for everyone else.

I draw these conclusions based in large part on the achievements of energy-efficiency policies that have been adopted in the United States since the 1970s:

1. Since 1976, the U.S. Congress, the Department of Energy, and the California Energy Commission, along with several other states, have set energy-efficiency standards for dozens of appliances and pieces of equipment, such as refrigerators, air-conditioners, and lighting equipment. The predicted net economic benefits of these standards are estimated conservatively at over one trillion dollars.[1] And, as will be shown, the actual benefits include product improvements whose value has not been counted in this estimate. Moreover the actual costs are much lower than were predicted. In many cases they were zero

or even negative (in other words the cost of the efficient product was even less than the cost of the product they replaced).

2. Energy-efficiency standards for new buildings, implemented in the United States by states, have generated at least $200 billion of net energy benefits. And, as I will describe below, the nonenergy benefits greatly exceed the energy benefits.

These immense savings are only the tip of the iceberg. The United States has not been consistent or aggressive in its efforts to promote energy efficiency. Policies that have clear benefit have been ignored as the issue became mired in political squabbles that were based on larger and, as I will show, irrelevant geopolitical or ideological debates. So the potential savings are much larger—well into the tens of trillions of dollars. These savings will generate millions of new jobs. The subject of energy efficiency is introduced in Chapter 2, and the issue is explored in depth throughout Part 1.

3. The benefits of environmental policy in promoting growth are not limited to energy efficiency, however. The Congress and the Environmental Protection Agency, along with the California Air Resources Board, have regulated air pollution emissions since the 1970s. The net economic benefits of the federal regulations have been estimated at $1 trillion, with $1.2 trillion of benefits being obtained for $220 million of costs.[2]

Of course, not all environmental policies are alike. Some work better than others and there are undoubtedly a few areas where environmental quality might come at the expense of some level of economic growth. But there is increasing evidence that well-designed environmental policies, including regulations, promote economic growth, and indeed are one of the few strategies that we understand that can do so.

The observation that well-designed environmental policies promote economic growth is particularly interesting because up until now, they have done so without trying. That is, economic development has not been a design objective of environmental policy: the economic benefit comes almost by accident. Therefore it is likely that, if economic development were an explicit goal of environmental policy making, we could do an even better job.

If the success of environmentalism leads to greater growth, why would anyone be opposed? This book explores how the politics of environmental policy are affecting the debate more than real economic interests. The politics begin with the organized opposition of the business community to almost all of the environmental initiatives that promise real change.

Organized Business Opposition to Environmental Protection

The relationship between environmental protection and economic development, not only in the United States but also throughout the world, has profound consequences for environmental policy. If environmental protection truly promotes economic development the business community should be supportive of environmental policies on the whole—even if a few companies oppose selected environmental protection plans. Instead (in practice) virtually all of the organized business community opposes serious environmental protection proposals.

Such opposition to environmental protection is particularly odd in a supposedly competitive economy. Regardless of whether environmental policy helps or hurts business in general, any particular environmental policy is likely to have winners and losers in different companies and in different industries.

For example, a regulation that requires increased efficiency for air-conditioners might be opposed by power companies with excess capacity for electricity generation that they sell at high profit to other utilities that are short of power (due to high air-conditioner usage), yet utilities facing the reverse situation should support the regulation. This regulation might be attractive to aluminum and copper manufacturers because it would require higher production of these materials to make more efficient air-conditioners. Naturally small businesses and building owners, whose electric bills would decrease if efficient air-conditioners reduced the risk of blackouts and kept electricity prices down, should support such efficiency regulation.

Surprisingly, however, this almost never happens. Instead the organized business community tends overwhelmingly to support the interests of those corporations who expect to lose from the policy change and ignores the interests of the potential winners. The self-expected losers, by and large, are economic incumbents—a relatively few large and well-established corporations that have sizable shares of their markets.

These economic incumbents are the big companies that think (or worry that) they have the most to lose from increased competition or technological innovation. They assume that they're defending their real economic self-interests by attempts to sustain their entrenched positions through political power—but often this is not a realistic assumption. Often large corporations fade into insignificance when they fail to adjust to new realities of unanticipated competition and technological progress. In fact, they often have the most to gain from environmental policies that require innovation because large, established companies may find it easier to develop and sell new technologies than their competitors.

Given this misperception, it's no surprise that economic incumbents might lobby against policies that change the status quo. What is a surprise is that the rest of the business community—and political leaders who support economic growth and competition—would join them.

Yet such political alliances are almost universal. Corporations that fear the consequences of a particular environmental policy will lobby vigorously to protect the status quo, while those businesses that would benefit remain silent—or worse, speak out in support of the self-perceived losers.

Changing Corporate Attitudes Toward Environmental Regulation and Protection

A recognition that environmental protection promotes economic growth could change the dynamic between corporations and environmental advocates and regulators. If environmental policymakers better understood the diverse interests of different corporations, government could develop policies that not only protected the environment more thoroughly, but also were more supportive of economic growth. A similar understanding could also transform the attitudes of environmental advocates. Simply put, if we all understood that less pollution means more profits for corporate America, environmental/business partnerships would undoubtedly be significantly more common and a more substantial part of the agenda of environmental organizations.

The fact that the business community and the environmental community have differing perspectives suggests that basic strategic decisions result from obsolete, knee-jerk reactions rather than rational assessments of self-interest.

I saw this in part of my work as an environmental advocate in the 1970s and 1980s when the positions of both the environmental and the business community were quite rigid. Environmental advocates supported enhancements in energy-efficiency regulations for products that used significant amounts of energy and, naturally, their builders and manufacturers opposed these enhancements. Neither side talked to each other.

The standoff began to change in the 1980s, after face-to-face negotiations over federal-level appliance efficiency legislation, and a dialogue emerged between some companies and environmental advocates. On occasion the companies that saw themselves as winners from a proposed regulation began to work in alliance with environmental advocates in support of that regulation. Friendlier relationships paved the way toward greater cooperation. I recall when an appliance company that had previously worked collaboratively with the efficiency advocates decided that it was going to oppose their position on a particular regulation—one that was open to public comment—its representative phoned me. Here's an excerpt of that conversation:

"I'm calling to give you a heads-up on the issue of the proposed efficiency standards for our product," the vice president in charge of governmental relations told me. "Just so you aren't surprised, I wanted to tell you that we are going to be on opposite sides on this particular issue."

"What is your opposition based on?" I asked him. "Wouldn't your company make more money with the regulation than without it?"

"More money? I don't know if we have done a calculation. Let me get back to you on that."

A few weeks later, the same industry spokesman called back and said, "The company thought about the issues a little more and has changed its position. We will now support the regulation."

"Great," I said. "Why such a reversal?"

"Well… our engineers and managers did a spreadsheet on the difference in our profitability with or without the regulation and found that their business would be more profitable with the higher standard."

"Okay," I continued. "So let me ask you a question: do you understand your competitors' businesses enough to be able to predict whether, if they performed the same spreadsheet, they would get the same answer?"

"Yeah, we know the competition would reach the same conclusion if they ran the numbers."

Yet none of the other manufacturers supported the rule in the end. Thus it is most likely that one of the following two things happened: either the other companies never analyzed how the regulation would affect their profit, but reacted automatically in opposition out of habit; or that other companies preferred to maintain a consistent business position of opposition to environmental regulations even if it hurt their own interests.

Such an observation raises deep questions about how the political/economic system truly works. How do corporate leaders decide about environmental issues, both those that affect their companies directly and those that affect them indirectly? How do the business and environmental communities affect the structure of markets and competition in the United States? What are the motivations for these communities' positions?

The Influence of Myth in Environmental Policy Debates

The crucial influence that myths and misunderstandings play in our national debate on environmental policy is a major subject of this book. I assert that myths—idealized and oversimplified stories—about how the economy functions and what are the appropriate positions for businesses or environmental advocates to take explain the entrenched positions of many large corporations on major environmental issues much better than any rational accounting of self-interest. As we debunk such myths, we open up major opportunities for partnership between businesses and environmentalists.

These myths also appear to have a decisive influence on the thought processes of our elected officials and public administrators, which increases the perceived level of controversy of environmental decisions and polarizes many of the political debates.

Debunking the Myths of Economic Theory

A key area of mythology focuses on economic theory. Throughout the last forty years, opponents of environmental protection have based most of their arguments on economics and, in particular, on the argument that government regulations are unnecessary in a competitive market and inevitably reduce economic welfare.

Mainstream economists seldom seriously address such an argument. However, careful examination of economic theory—before we even look at economic data—shows that the concept of a conflict between regulation and market forces (the idea that regulation is the opposite of market forces) is fatally flawed. Instead regulations are a necessary part of a functional free market. (I address this issue throughout the book.)

The supposed political battle between the advocates of market forces and the advocates of regulation may actually serve as the basis of business opposition to environmental protection; or it may be just an advocacy technique used when it is thought to be politically effective or convenient. But the explanation that perhaps it is just a tool of advocacy begs the question of what might be the true motivation of its advocates.

Do most businesses truly see regulation as an attack on free enterprise? If so, what is the basis for this myth's acceptance? And why do they support, through membership fees and other expenses, such private-sector organizations as trade associations whose primary purpose often is to promulgate and enforce regulations? This expense can be considerable. One major trade association employs almost half of its staff on the development and certification of industry standards. Surely its members don't consider this activity something that conflicts with free markets!

Myths and Political Alliances

The myth that in order to support free markets one must oppose regulation is only one of a number of myths that are preventing a constructive dialogue between business and environmental leaders—a dialogue that could lead to policies that more effectively promote economic development while they also encourage environmental progress.

For both environmentalists and their opponents, many of the myths that block constructive dialogue focus on the supposed alliances of the opposition. Many of the arguments on both sides address issues far removed from the environment, but closely connected with what each side believes the other really desires.

I present throughout this book one important red herring that afflicts the arguments of both sides—the issue of government-planned economic activity, as opposed to market-based economic activity. Many of the arguments against environmental protection actually are concerned more about the issue of potentially dangerous top-down government economic planning—the sort that was used in the Soviet Union—than they are about the environment itself.

The question of government-planned economic activity is more important than it appears because much of the opposition to environmental policies disproportionately emphasizes the issue of government control versus individual choice as the primary reason to oppose environmental protection. A surprisingly large percentage of the anti-environmental writings on the Internet address the authors' opposition to a state-controlled economy and their apparent belief that environmentalism is a stalking horse for big government. This is all the more surprising given that Communism is no longer an ideologically competitive system since the fall of the Soviet Union.

Perhaps even more difficult to accept is how anti-environmentalists can ignore the fact that Communist countries had the world's worst record on pollution control and environmental health. Why would environmental advocates want to promote a system that so obviously failed?

Anti-environmentalists' attempts to equate environmental protection with a command-economy that overturns free markets are ironic, moreover, because the evidence suggests the reverse: environmental protection policies tend to create or strengthen competitive markets. Yet, equally inexplicably, environmental advocates seldom use pro-market arguments to support their recommendations. Instead environmental writings often emphasize images of big, bad corporations that intentionally destroy the environment just to increase profits. These myths are intended to appeal to a populist and anti-corporate sentiment that may be a defensible political position but that has little to do with the environment.

Environmental mythology often focuses on the identification of supposed bad guys who are responsible for pollution. Yet to frame the problem this way, even if or when it is correct, leads in unproductive directions. It sets up the debate as a power struggle rather than a policy decision, with business on one side and the public on the other.

To frame environmental issues as a struggle between corporations and the public interest just reinforces the myth that environmentalism is a stalking horse for government control of business. It also encourages an already-strong tendency for businesses to react to issues based on peer identification and group loyalty rather than on self-interest and competitive advantage.

The ultimate irony is that when business leaders respond to environmental issues with a unified voice of opposition, it simply reinforces trends toward the very result that anti-environmentalists fear most: top-down economic decision-making by political forces rather than market forces. Free-market advocates should not care whether markets are hobbled by government's or business's politically chosen policies: both are in fundamental conflict with the concept of competition.

Environmental Protection Can Overcome Failures in the Marketplace

The American economy is beset by deep and widespread market failures that stifle innovation and competition and thus reduce economic growth. Mainstream economists have ignored most of these problems, perhaps because they are not readily apparent in analyzing markets that are stable and that don't involve rapid technological change. But they are at the heart of analysis and policy development in the environmental area, and particularly in energy efficiency.

If we identify these failures and develop policies to correct them, we can make the actual economy function more closely to how an ideal economy functions. Many known and successfully tested ways to do this were discovered in the context of evaluating environmental policies, rather than in the context of studying economic development. More widespread recognition of these solutions and discussion about how to apply them more broadly would logically create win-win solutions to the related problems of how to protect the environment and accelerate economic growth.

How to Use This Book

Thus far we have a preliminary overview of the reasons why environmental protection policies promote growth. The track record of energy-efficiency policy forms the cornerstone of this argument, so the following chapters focus on how such policies have succeeded in reducing pollution and energy costs while saving money and creating jobs.

Three principal sections comprise the book's structure. I delve into the practical side of environmental protection in Part 1, addressing the direct and indirect effects of environmental policies that impact the economy. The focus is on energy efficiency, where the evidence is clearest and where the author's experience is concentrated.

Part 2 focuses on economic theory because so much of the opposition to environmental protection policy is based on a reading of economic theory that oversimplifies the actual economic arguments by ignoring the assumptions on which the theory is based. Indeed, most of the arguments against higher levels of environmental protection, whether through regulation or incentive, and regardless of whether the opposition is broad and general or targeted to one specific law or regulation, rely heavily on poorly thought out theory and on weak (or nonexistent) actual evidence. Part 2 examines why a more careful look at the theory leads to markedly different conclusions, and how a comparison of the naïve theory with facts (when they are available) almost always demonstrates that evidence refutes the simple theory.

Part 2 also offers a more complete form of economic theory, one that is based on real-world experience with how markets work, again with a focus on energy and the environment. This section explores a number of ways that markets fail to produce the best outcome that the simpler theory says they should. Some of these ways relate to the structure of markets, some relate to how humans behave, and others relate to the political forces that shape the real economy. Understanding how markets fail to deliver what they should is the first step towards solving the problems and increasing growth.

Part 3 presents the politics and ideology of environmental controversies, with an emphasis on myths that both pro- and anti-environmental activists rely on to formulate and sell their positions. The examination of such political problems, as well as the structural problems described in Part 2 is essential to create a pro-environment, pro-growth strategy.

Part I

Energy Efficiency and the Economy

I introduce in this first section the evidence and the mechanisms for how and why environmental protection policy enhances economic growth. The focus here is on practical experience and real-world evidence rather than theory. (Theoretical issues are explored in Part 2.)

Chapter 1 provides a basic explanation of energy efficiency—what it means, how important it is to the economy, and how it serves as a model for how environmental protection affects the economy. This chapter also provides a more detailed overview of the main ideas of the book, all of which will be developed further in the following chapters.

I discuss in Chapter 2 the direct ways that environmental policy promotes economic growth. I illustrate how the measurable economic benefits of environmental protection are almost always larger than the costs.

For example, the benefits from energy-efficiency measures that have already been undertaken in the American economy exceed $1 trillion, and the benefits of the Clean Air Act alone are also valued at over $1 trillion by U.S. Environmental Protection Agency and Office of Management and Budget studies. These two policies alone produce benefits valued at 20 percent of a whole year's output of the American economy.

Decisions that produce larger benefits than costs promote economic growth, by increasing the efficiency of the economy. They also create new jobs.

The observation that environmental protection is beneficial economically is neither a new nor original idea—papers that demonstrate this fact were published as early as the mid-1970s. It is well-known and generally well-accepted among energy agencies and governmental agencies responsible for environmental protection. Yet business leaders and policymakers still lack a solid understanding, so it merits a detailed discussion here.

Another way to express this proposition is to observe that the economy is full of unexploited business opportunities to reduce energy use or otherwise improve environmental performance at a profit. This is both a business opportunity that corporations can implement on their own and one that they can implement through advocacy and organization.

The straightforward opportunity is for business managers to examine their own operations and look for ways to improve efficiency and profits. If they do so, their actions will create opportunities for the broader economy—for other businesses that sell the efficiency services or the equipment needed to enhance efficiency or reduce pollution.

The other opportunity is for businesses to advocate for government policies that will help them to identify the opportunities and help to finance them, or to create new markets for businesses that sell efficiency/environmental services or products.

A close examination of what happens when efficiency policies or other environmental policies are actually implemented shows that they usually produce unanticipated benefits or side effects that increase the benefits dramatically. Chapter 3 explores how the indirect benefits of environmental protection policies may be even larger than the direct benefits. This results in what I call the "strong form" of the growth proposition: that environmental protection policies promote innovation and creativity as affected companies find new and better ways to comply with environmental performance goals, and that these innovations generally produce gains that are unrelated to the expected improvements—and usually much larger. These additional growth benefits are not even limited to the firms that comply with the regulations or incentives—they can occur in other parts of the economy as well.

The strong form has been discussed in a limited way in a few academic reports, but is largely unknown to a broader business and policy audience. Even in its academic form, the strong form of the growth proposition is not well elaborated or explained, either in terms of the evidence that supports it or the mechanisms by which it can work. Chapter 3 provides the first steps in such an explanation.

1 | The Critical Role of Energy Efficiency in the Economy

Energy use is at the heart of many of America's (and the world's) most critical environmental, geopolitical, and economic problems:

- Over 80 percent of America's greenhouse gas emissions—pollution that causes global climate change—is caused directly by energy consumption.

- At least 30 percent, and as much as 90 percent, of conventional air pollution in the United States results from energy use: the emissions of sulfur oxides (SOx), nitrogen oxides (NOx), volatile organic chemicals (VOCs), carbon monoxide (CO), and small particles that are collectively responsible for more than 30,000 deaths annually in the United States, and well over a million worldwide.[1]

- The demand for energy prompts proposals to drill for oil and gas or mine for coal in environmentally sensitive areas—and areas that currently are not industrialized or even well connected culturally to the industrial world. These land-use conflicts promote such social ills as opportunistic dictatorships, civil wars, and vast discrepancies between the rich and poor, in addition to environmental problems.

- Transportation and conversion of energy creates added environmental problems ranging from oil spills at sea to land-use conflicts over transmission and pipeline rights of way.

- High worldwide demand for oil directs large sums of money to politically unstable and undemocratic regimes, some of which support terrorism. "[America is] in a war. It is a war against open societies mounted by Islamo-fascists, who are nurtured by mosques, charities and madrasas preaching … intolerance… and financed by medieval regimes sustained by our oil purchases."[2]

Energy costs account for an estimated 6 to 10 percent of the U.S. economy, depending on oil and gas prices. And increases in energy prices have twice triggered economy-wide inflation problems in the 1970s, not just in America but also globally.

Energy is imported into most regions and into the United States as a whole. The cost to the U.S. economy from imported energy has exceeded $100 billion annually every year since 2000 and is on track to exceed $200 billion in 2006. This cost reduces domestic employment and economic growth by redirecting financial resources abroad rather than supporting our local or domestic economy.

A reduction in energy use, therefore, clearly provides significant environmental, economic, and security benefits to the U.S. economy.

How can we reduce our energy use? Most people are already familiar with how to make lifestyle sacrifices and compromises that can cut energy consumption: reduce temperature settings on thermostats in the winter and raise them in the summer; avoid unnecessary driving; be vigilant about turning off unneeded lights; and don't run the dishwasher until it's fully loaded. Yet such efforts are often unappealing to many people, and no government policy exists that is effective at encouraging adherence to them.[3] What then is a practical and realistic way to do this?

Energy Use Reduction

Fortunately there are more effective ways to reduce energy usage. Unlike many products, energy isn't desired for its own sake. Energy is merely a means to an end. We call the end goals energy services: they include comfortably lit offices or homes, cool temperatures indoors in the summer, the ability to travel conveniently, and the ability to clean clothes.

A number of different technologies can provide energy services, and different choices produce different levels of energy demand for exactly the same energy service. Serious study of energy services began in the 1970s. Analysts used the term energy efficiency to mean reductions in energy use that maintained the level of energy services; in contrast, conservation meant reduced energy use that involved sacrifices or compromises in amenity.

Energy Efficiency: A Good Investment

Energy-efficiency policies provide better energy services with less energy usage because they promote the substitution of improved technologies and designs.

Consider energy efficiency actions as investments. An improvement in technology or design typically costs more money—at least at first. The additional cost is paid back over time by the value of energy savings, just as an initial investment in a building is paid back to the investor by the rent he or she receives over time. Chapter 2 introduces numerous examples of energy-efficiency measures that are superior investments to most or all of the options available in financial markets, including stocks, bonds, commodities, and real estate. Common examples include improved insulation in buildings, light-emitting-diode-based traffic signals, clothes washers that eliminate the need to drag clothes through a full tub of hot water, and hybrid-drive automobiles that use less gasoline than conventional cars.

Opportunities for investments in energy efficiency at surprisingly high rates of return abound throughout the economy. It is easy to reduce energy consumption by at least 20 to 30 percent in an existing commercial building; and new buildings can cut their energy demand by 50 percent or more. Technologies for leak-free heating and cooling ducts in homes, along with heat-rejecting windows and better insulation, can reduce energy usage by 30 to 50 percent. Additionally, high-efficiency appliances and equipment in homes and commercial buildings can reduce energy use by 50 percent or more.

Similar opportunities are widespread within the industrial sector as well.

Policies that can promote the acceptance of such energy-efficiency products and designs have obvious and remarkable economic benefits. Efficiency measures installed in commercial buildings typically pay back their initial costs in two to three years. But as the initial, one-time additional costs provide continuing returns over time, they actually are long-term, profit-generating investments because they cut costs year after year. (This viewpoint provides a comparison that is more appropriate in terms of economics or business management.)

Take, for example, that two-or-three-year payback for a commercial building efficiency project. The cash flow on the project is equivalent to a return on investment of 30 to 50 percent per year. Home energy-efficiency investments typically offer an annual 10 percent return or better. Efficiency upgrades in industrial plants as well as commercial buildings provide a 30 to 50 percent return or better.

Consumers and businesses who invest in energy-efficiency opportunities at 10 to 50 percent (or better) rates of return are obviously making more money than if they had invested in conventional options that typically return 5 to 6 percent per year (after inflation) at best. A long-term investment in the stocks of U.S. companies has returned about 6 percent annually for almost a hundred years, but this figure diminishes with added corrections for taxes and trading commissions. Long-term investments in bonds yield 1 to 3 percent per year (after inflation.)

The most comprehensive studies indicate the potential benefits to the U.S. economy for making money from energy-efficiency investments are in the trillions of dollars. The benefits evaluated include only the direct value of energy savings to the consumer. In fact, the economy as a whole does even better because lower demands for energy result in lower energy prices for everyone, not just for the individual or business that makes the efficiency investment.

For many forms of energy, small changes in supply or demand can lead to large changes in price. The U.S. markets for natural gas have illustrated this characteristic since about 2000, and more recently world oil prices fluctuate strongly based on even small changes in supply or demand. In such cases, energy savings lead to two economic benefits: (1) the person who saves the energy clearly saves money on an individual basis because less energy needs to be purchased; (2) yet everyone else saves money too because the first person's efficiency investment reduces the overall demand for energy, thus reducing the price. All others will save money from lower energy prices even if their consumption remains the same.

The economic benefits from such savings can be substantial. In one important case, a piece of energy-efficiency legislation for encouraging efficient buildings and equipment considered in the 109th Congress (parts of which the Energy Policy Act of 2005 incorporates), the direct savings to consumers who owned the more efficient buildings and equipment was estimated at $40 billion per year by 2015.[4] For everyone else, the benefit of the lowered natural gas prices was estimated at over $30 billion per year.

Also, energy that is not consumed typically displaces energy that would have been imported from outside the United States or would have been produced in remote areas where there aren't many unemployed workers who can benefit from the increased job creation. In contrast, energy-efficiency investments go largely to pay for existing factories or in the construction industry where jobs are needed.

This argument is equally valid for any other country or region that is a net importer of energy.

Energy efficiency creates jobs in several ways: first, the production and installation of energy-efficient products requires workers; second, because efficiency investments save money, consumers and businesses will utilize savings for alternative purposes, expanding employment throughout the economy. (A few jobs will be lost in the energy-supply sector, but energy supply provides only about half as many jobs as the rest of the economy per dollar of revenues.) Third, energy efficiency reduces the demand for energy, causing prices to drop. Lower energy prices allow lower interest rates and higher growth.

All these benefits occur before we even consider the direct economic advantages of the environmental improvements themselves, which can be very significant. As noted earlier, the U.S. Environmental Protection Agency estimated that the benefits of the Clean Air Act exceed $1.2 trillion; these benefits primarily occur in the form of reduced medical expenses and reduced deaths due to cleaner air. The benefits overwhelmingly outweigh the costs of compliance, which were estimated at $220 billion. As energy efficiency also reduces air pollution, such a reduction increases clean-air savings.

The federal Office of Management and Budget under the George W. Bush Administration provided a broader and more recent analysis of environmental regulation that reached a similar conclusion. It estimated that the economic benefits of major U.S. Environmental Protection Agency regulations amounted to between $120 and $193 billion annually, while the cost of compliance was only $36 to $43 billion annually.[5] Costs often are systematically overestimated (see Chapter 3).

Such are some of the examples of the direct economic growth proposition: that the direct and obvious benefits of the environmental improvement outweigh the costs. Both the academic world and government have discussed this for over thirty years, yet the message still appears slow in its arrival—a continued communications problem. (See also Chapter 10.)

Energy-Efficiency Policy and Economic Growth

As dramatic as the results of the direct version of the economic-growth proposition are, accumulated evidence and experience over the past ten years provides the basis for a strong version of the proposition: that environmental protection affirmatively enhances economic growth... even more effectively through its indirect consequences.

The strong version is based on the practical observation that when environmental regulations and incentives have been particularly sustained and aggressive, they have spurred innovation in industry that led to overall reductions in cost and improvements in areas of production unrelated to the actual environmental goal.

Innovations in technology and savings in the costs of inputs such as energy increase the overall productivity of the economy. And increasing productivity creates more jobs...and higher-paying jobs. For energy savings, the employment benefits are particularly strong because the energy supply sector provides few jobs per million dollars of revenue. Even diverting expenditures from energy supply to efficiency without saving money would increase employment; saving energy at a net economic gain produces even more jobs.

In the energy-efficiency area, the indirect economic benefits take two forms:

1. Nonenergy benefits
2. Cost reductions (for the products that incorporate efficiency technology)

Nonenergy Benefits

For a large number of energy uses (e.g., washing clothes, lighting a factory or an office, and so forth), the more efficient option produces benefits in nonenergy related areas that far outweigh the cost savings from reduced energy bills. For example, more efficient lighting in commercial buildings, including the use of natural daylight, improves how the building functions. Retail sales increase in day-lit areas of stores, and students in day-lit classrooms have noticeably higher test scores. Certain energy-efficiency technologies have produced a measured benefit on the productivity of office workers. Even the small human productivity benefit observed in the studies (in terms of percent improvement) is worth a great deal more than the direct value of energy savings.

New clothes washers (identified by the EnergyStar® label in the United States) use technologies that either sprinkle water on the clothes or drop the clothes into a shallow pool of water at the bottom of the tub. The new designs replace traditional technologies that require washers to use a full tub of hot water and an agitator to drag clothes through the water. Such innovations reduce water consumption by over half and energy consumption by two-thirds or more. Reductions in the cost of water consumption and in the cost of energy required to heat the water pay for the improved efficiency in about three years. Additionally, the more efficient washers require less detergent—a cost savings that can even surpass the value of the overall energy savings.

Yet nonenergy benefits overshadow all such direct energy or energy-related benefits. Research by the washer manufacturers has documented that their efficient products both extend fabric life and clean clothes better. Marketing experts for the major manufacturers claim consumers are much more likely to buy the EnergyStar® washers because they improve cleaning ability and enhance fabric care than due to saving energy, water, or detergent.

Light-emitting diodes (LEDs) are used for traffic signals because the technology is ten times more efficient than incandescent lights in providing red, amber, and green lighting. Notwithstanding the energy savings, which pay for the extra costs of the signal head, the bulk of the savings for municipalities that install these signal heads is because of the superior LED lifetime, which leads to savings in the cost of maintenance staff time that exceeds the value of the energy savings.

However, the actual benefits are even better because an LED red light contains dozens of individual LEDs and it is virtually impossible that all will fail at once. So if a conventional traffic light burns out, the driver sees a blank space where a red light should be, but if one or several LEDs burn out, the (somewhat irregular) red light remains functional, and the potential for hazard is avoided.

An additional benefit of LED traffic signals is that their energy consumption is so low that they can be backed up by battery power in the event of an electricity failure.

In these and an increasingly large number of other cases, the nonenergy benefits exceed the energy benefits; thus, an expected 30-percent return on investment may actually be a 100 percent—or even 1,000 percent—return on investment! (See Chapter 3 for additional examples.)

Yet efficiency can produce even more benefits.

Environmental Policies Generate Cost Reductions

In a number of cases, follow-up studies measuring the actual costs of energy efficiency compared to the projected costs (on which the rates of return were based) indicate that the additional costs are significantly less than predicted, and are zero or even negative in many cases.

For example, in 1992 when manufacturers of central air conditioners first had to meet strict nationwide standards for air-conditioner efficiency (they initially opposed such regulation, but later supported it, as it turned out), the average price of an airconditioner did not increase by $350, as the government projected—nor by $700, as manufacturers predicted—but actually failed to show an increase at all!

Today, the air conditioner industry is completing a transition to a 2006 standard that is 30 percent higher in measured and tested efficiency than the 1992 standard. By early 2005 the trade association for air conditioner manufacturers announced that its members were making great progress toward meeting the standard without undue disruption or cost. (No specific predictions on cost could be made because of anti-trust law considerations.) However, some quantitative information is beginning to appear that depicts the nonenergy benefits of the new standard's encouragement of innovation.

Most of the analysis the Department of Energy considered showed that the new standards would result in larger and heavier airconditioners. But in August 2005, one of the largest manufacturers announced a line of 2006 air-conditioners that would be 20 percent smaller and 30 percent lighter than the units the DOE had considered. It also would use a totally non-ozone-depleting refrigerant, and at a 40 percent reduced quantity at that.[6]

Similar experience was recorded within the refrigerator industry, not once, but repeatedly, following energy regulations in 1977, 1979, 1987, 1990, 1993, and 2001, in addition to ozone-depleter phase-out requirements made effective circa 1995 and 2001. The price of a refrigerator in the United States has declined steadily, after adjustment for inflation, by almost 50 percent over the past thirty years.

I've found that when a manufacturer must redesign a production process for an environmental goal, such as energy efficiency—either because of regulation or in response to strong economic incentives—the company uses such an opportunity to redesign for a number of other production issues that can cut costs or improve quality. These energy-efficiency improvements might cost extra money if they had been undertaken on their own, but this is rarely the case.

For example, when the executive director of a national energy-efficiency organization contacted a transformer manufacturer about its ability to produce a new design that would be much more energy efficient, the manufacturer responded that if the program was aggressive enough to assure a market for the new efficient product, the company could produce it. When asked how much extra the efficient product would cost, the manufacturer responded that he would sell it for the same price. How could this be done? Because when the manufacturer built the assembly line for the new transformer, it could readjust other factors of production besides energy efficiency and pay back all its incremental costs from in-plant savings. The company would not make all the other changes by themselves (without the energy efficiency), but would undertake the entire project only if there was a market advantage to the energy-efficient product. Again, the energy improvements might have added to the cost if they were undertaken on their own, but this doesn't often happen in the real world; in real life, the policies that promoted the energy-efficient transformer were the only way to start the entire project, and in its entirety the project produced the efficiency gains for free.

Similar results characterize the continually advancing requirements for vehicular air-pollution control: no increases in car prices have resulted due to the upgraded stringency of environmental requirements. Automobiles have been subject to an increasing number of regulatory requirements since the 1970s. In additional to national safety requirements, first for seat belts and then for air bags—and national fuel economy standards— the California Air Resources Board has imposed increasingly tightened emissions limits. While the first round of emissions standards in the 1970s imposed an additional cost per car, all of the subsequent improvements were obtained without any incremental cost increase. A retrospective study noted: "Regulations to improve vehicle safety and environmental and energy performance do impose additional costs, but these costs are neither permanent nor cumulative."[7] In other words, all environment regulations imposed after 1980 achieved their smog-reducing benefits for free.

In short, the need to comply with the environmental policy helps shake industry out of complacency and encourages more innovation and competition.

How is it that complying with environmental regulations for energy efficiency encourages competition and innovation? Evidently, the U.S. economy isn't as competitive as naïve economic theory suggests. Policy makers can enhance innovation, competition, and economic growth by identifying those areas where competition becomes subdued and limited, and where manufacturers have no real reason to produce something differently today than it has been produced in the past.

Opposition to Energy Efficiency

Surprisingly many areas can be found where competition is subdued and technologies do not advance despite the potential for high-payoff improvements. Water heaters, for example, subject to relatively little environmental regulation following World War II through the end of the century, have remained essentially unchanged in design and performance. Efficiency levels of gas-fired water heaters actually declined from the 1940s until they were subject to weak regulation in the 1970s. Similarly, refrigerator efficiency declined during this same time by almost 50 percent until they were regulated in the 70s.

Many examples indicate where industry is stuck in a pattern that requires little or no innovation. Many of these cases are marked by situations wherein a relatively small number of companies shares the market more or less equally, and wherein market shares have not shifted dramatically in recent history. Restoring competition to these industries by introducing new technology imperatives provides a successful explanation for how environmental protection drives economic growth.

Business executives do not necessarily find being in a stable industry—one with only slow changes, if any, in technology and markets—an uncomfortable place. It is not difficult to imagine that the government relations departments of companies in such an industry might argue vigorously against environmental policies that might upset the apple cart and provide openings for competitors to take away market share. This would be true even if the most likely scenario is that the established company maintains its dominance and just sells a better and perhaps more expensive product at a somewhat higher profit.

Much of the opposition to environmental-protection policies is better understood as protection of economic incumbency by established interests rather than the promotion of economic growth or the interest of business.

Economic incumbency is used here to denote those corporations that have a strong, established position in the marketplace that they are attempting to maintain through lobbying or other political activity. Not all corporations, or even all large corporations, are necessarily economic incumbents. Some markets inherently force competition on a level playing field even when one or several of the participants are large or old.

For example, the computer chip industry continues to introduce higher-technology products with ever-decreasing prices despite its concentration. We also see innovations in aircraft design regardless of the fact only two firms, Airbus and Boeing, both old and well-established, dominate the global market.

The Political Abuses of Economic Theory

The motivation to protect incumbent economic interests from competition is easy to disguise as being pro-business or pro-market because the simplest reading of economic theory provides a strong basis for leaving the status quo alone. Opponents of environmental policy and regulation in particular almost invariably invoke economic theory as a reason for their opposition. Of course, the proposition that environmental protection enhances economic growth is contrary to naïve economic theory.

In its simplest form, economic theory suggests that a free competitive market produces the best possible outcome for everyone, given the initial conditions and a number of assumptions. According to this theory, any government interference in the marketplace can only make things worse. (Chapter 4 concentrates on a number of detailed problems with this theory.) Yet perhaps the most fundamental error that those who try to oppose regulation to the so-called free market make is the implicit and unstated assumption that what we already have is an ideal free market.

Yet to the extent that the market is dominated by a limited number of firms, and the regulations by which industry works are written in industry forums where large companies can more or less dictate the results, the status quo already departs from what economic theory would predict. (Chapters 4 and 6 address this further.) And the extent to which large market size leads to special deals with other businesses, or that consumers or users don't have the means to evaluate the energy-cost or environmental implications of their choices, we don't have a free market in the first place.

For example, one of the barriers to the availability of compact fluorescent lights is that shelf space at supermarkets is assigned to large suppliers based on deals they negotiated with the supermarket chain; thus, the incandescent lamp manufacturers kept out manufacturers of compact fluorescent lamps by reducing the amount of shelf availability.

The example of shelf space allocation without regard to whether or not consumers buy the products is definitely not how a free market is supposed to work; deals in free markets are transparent and all consumers are supposed to have access to the same choices of goods and services to purchase.

So if our starting point is not an ideal free market, and if many of the assumptions necessary for markets to be free are not fulfilled, government-induced changes are not inherently either beneficial or harmful, but instead have to be evaluated solely on their merits.

Economic Fundamentalism

If we examine the websites of anti-environmental advocacy organizations and think tanks, and look at the testimony of industries and trade associations opposed to environmental policies, we see that they hew to an oversimplified and ideologically rigid form of economic theory. This is akin to a fundamentalist religious belief in that its strongest proponents aren't interested in facts, but rather claim to acknowledge a revealed truth.

Whatever one can argue about the validity of revelation with respect to religious concepts, revealed truth has no place in discussions of environmental policy. Indeed, economics is accepted as a science, meaning a field of study in which facts are essential to validate (or disprove) theories.

Surprisingly little discussion or academic analysis focuses on what the facts are concerning how well markets work on environmental qualities in general, or even energy efficiency in particular. But repeated studies in the regulatory and policy forums show the same results: that the simple economic theory—that markets without government involvement produce the best results for everyone—is inconsistent with the data.

Multi-million-dollar evaluations of billion-dollar utility-sponsored programs that encourage their customers to invest in energy efficiency have been debated vigorously in front of Public Utilities Commissions and Public Service Commissions throughout the nation wherever utilities were charged with spending customer money to produce customer benefit. Many state officials, as well as consumer-advocacy organizations, are highly loath to provide this money to utilities without overwhelming evidence that such programs produce public benefit.

Time after time, in such adversarial formats, Commissions have ruled that the evidence provided by the utilities supports their claims that large, unexploited opportunities for energy efficiency exist among their customers and would not have been realized but for the intervention of the utility.[8]

These unrealized opportunities for cost-effective investments in energy efficiency are so large that when California faced an electricity crisis in 2001 and quadrupled its spending on energy efficiency from $250 million to over $1 billion, not only did it continue to find more energy-efficiency potential, but it even obtained four times as much savings as in previous years, at the same cost per kilowatt hour saved as it had been experiencing in past years. In other words, there were no diminishing returns, as would normally be expected by economists.

More broadly, the proposition that what we have now is a free market is contradicted in many cases by obvious facts; in some cases the less-obvious details are even more strongly in conflict with the proposition that the status quo approximates and ideal free market.

Consider, for example, electric utilities. In most parts of the

United States, utilities, or at least the utilities that sell electricity to the consumer, are regulated monopolies. Competition, within states that allow it at all, is limited to the competition between generators of electricity—the owners of power plants. Good public policy, perhaps, but it is nothing like an ideal free market.

Worse yet for those who believe that free markets create the greatest prosperity for everyone, these utilities are often regulated in a way that increases profitability to the extent that the regulated utility encourages uneconomic uses of energy. Thus, increasing profits for one industry come at the expense of profits and growth for everyone else. (See additional details in Chapter 11.) The main point is that pro-environmental regulation that makes utilities more profitable to the extent that they increase their customers' efficiency provides more economic benefits than the current situation. Reformed regulation can provide new business opportunities for the utility company (selling efficiency) as well as aligning its business interests with those of its customers.

Despite indisputable evidence to the contrary, the attraction of naïve economic theory remains high. There are, for instance, a number of studies regarding the Kyoto Protocol, an international agreement establishing mandatory limits on the emissions of greenhouse gas pollution. This agreement, implementing the United Nations Framework Convention on Climate Change, went into effect in 2005 despite the refusal of the United States to participate. Most of the widely publicized governmental and academic studies that conclude complying with the Kyoto Protocol would be relatively high not only fail to consider the conclusions of studies that look at the potential for cost-effective energy efficiency, but also even fail to cite them. In other words, arguments in opposition to naïve economic theory are not even refuted, but rather are simply ignored. (See Chapter 10 for further discussion.)

Perhaps one reason for the strength of the simple economic explanation for the world is that it provides a potential alliance between economic fundamentalists who believe in this model and entrenched economic interests that stand to benefit from its acceptance in the public-policy arena. The most obviously disastrous consequence of such an alliance is evident in California's experience with restructuring—or so-called deregulation—of electricity in the late 1990s. (See Chapter 5.)

Establish More Competitive Markets

Despite the focus on weaknesses in the traditional economic model, most recommendations here are pro-market. Evidence allows for reasonable conclusions that markets generally work best as a way of making economic decisions, but that establishing the ground rules for markets so that they actually do work is much more important than previously believed.

Ground Rules and the Role of Regulations

These ground rules include the definition of property rights and other legal issues relating to what sorts of market activities are fair and what sorts of activities are illegal, and also in many cases writing regulations (often written by the private sector) that define what traded products and services really are.

Recognizing the role of regulation in the market can help enhance competition. (I discuss regulation in greater detail in Chapters 4 and 10.) Rather than being an imposition on markets, regulations are in fact a necessary component of free and competitive markets. Regulations establish what the characteristics of products and services are, allowing existing companies and new entrants to a market to compete by offering the same product—namely one that meets the requirements of the regulation—at the lowest cost.

Introduced at least as early as the 1800s, regulations are vital to the U.S. market: the American private sector maintains over 40,000 regulations overseen by over 400 standard-setting organizations.

The real issue of regulation is not whether or not they are needed, but rather who writes them. The danger to a competitive market is that regulations may be written by private organizations, dominated by economic incumbents, to protect their existing products from competition. The process of government regulation is in some cases more resistant to the pleas of economic incumbents and can encourage innovation. This is particularly true for environmental regulations that are intended to promote new, cleaner technologies.

Overcoming Market Failures

Even with fair and just rules, markets often fail to produce theorists' expected optimal outcomes. Typically these failures result from identifiable problems or realities related to how human beings function, both psychologically as individuals and socially in their interactions in groups. Economic as well as environmental outcomes can be improved if policy makers focus on the widespread failures of markets and recognize either how to overcome such failures or how policies can compensate for the results of the failures.

A substantial body of research exists that addresses so-called market barriers and market failures. Yet economic fundamentalism compromises even this work, to a greater or lesser degree. In the case of explorations into the ways that markets can fail, the fundamentalist belief is that unless we can explain the exact reasons why failure occurs, it doesn't actually exist (regardless of contradicting data); or that if we can identify and correct the failures, then markets will function perfectly.

However, some failures are more fundamental and inherently unfixable. In this case, we need policies that cause real markets to produce the results that theoretical markets would produce. These policies may not be limited to simple corrections of identified market barriers such as lack of information.

The four different terms below each describes a type of reason why real markets fail to work as theory dictates. In Chapter 6, the term market barriers describes relatively easy-to-understand problems that are also easy to fix. For example, a market barrier that prevents homeowners from investing in enough energy efficiency in their homes is the lack of information on efficiency. The next owner may not recognize that the home is efficient and therefore refuse to pay more for it. The simple solution is to provide home-energy ratings that disclose the value of efficiency.

Market failures is a term to describe more pervasive problems, such as the fact that we live in a mass-production economy. A new technology might be extremely attractive if it sold, say, for $100, which would be the case if it were produced at a rate of ten million per year. However, because it is new and only produced by the hundreds, it sells for $500 and cannot be marketed successfully.

An even more difficult class of problem is one I call human failures. The most important examples relate to people's aversion to taking risks. When corporate managers are risk-averse in their actions on behalf of the corporation, they are missing the opportunity to make extra profits for their shareholders. Aversion to risk may or may not be a reasonable personal choice, but it leads to the wrong answers when people act as agents for others.

Institutional failures also occur when companies that should be following competitively neutral rules written by a disinterested party get together with their erstwhile competitors to lobby the government to regulate in a desired way, or simply write the regulations themselves. Whatever the advantages or disadvantages of this process, it is a failure of the market that tends to reduce competitive forces.

Reduced failures lead to markets that work better. If we can understand the failures of the market directly or discover through trial and error that a certain policy correction works (even if the reason is unknown), then we can solve the problems of the many types of failures of the market.

Markets still don't solve all problems. Yet a pragmatic approach would appear to focus on the development of effective markets as the primary mechanism for economic decision-making, supplemented only when a known problem exists. Primacy is based on the observation that economies that rely heavily on market forces are perceived as successfully positioned to prosper. However, countries that focus on overcoming or compensating for market failures are even more successful.

The Fear of Government Control

Fear of government control by business is a key recurring theme when examining the politics of anti-environmentalism. Noted in the Introduction, this fear is often expressed in broad terms that relate to property rights, as if the main concern was that an environmental restriction would reduce the ability of landowners to sell their property for more money.

Yet looking more closely at the attacks on the environment, it's evident that the overriding concern about so-called government control is actually the fear of principles or ideas associated with centralized economic planning. Such an association is based either on the content of pro-environmental arguments or on the affiliations of the environmentalists. This connection influences or even dominates both conservatives' and business's written attitudes toward environmental protection.

Perhaps the most striking manifestation of the connection described above occurred in the 1990s when a number of prominent conservative writers produced articles that attacked the environmental community for its advocacy of mitigating global warming through such international agreements as the Kyoto Protocol. Many of the op-ed pieces denied the scientific reality of global climate change and pointed to purported evidence to the contrary.

Yet surprisingly each of these articles, which appeared independently written, concluded with the argument that even environmentalists didn't truly care about global climate change. What environmentalists actually care about, these authors asserted, was to increase government control, or world government control, over the lives of citizens and corporations. Environmental protection was just a ruse to cover the true goal of self-proclaimed environmentalists: global governmental control.

The statement is odd for several reasons: first, having personally worked in the environmental-advocacy community for over thirty years, virtually all persons, regardless of whether I agreed with their positions, were truly motivated primarily by environmental protection. Many of these individuals hold strong political beliefs on other subjects as well, but the environmental policies are designed to solve environmental problems and not to address other hidden agendas.

Second, this argument seems to suggest an odd logical flow:
1. Increased government control is bad
2. If global climate change is a problem, it will require increased government control
3. Therefore, global climate change couldn't truly be happening.

More broadly, both sides of the debate, to a greater or lesser extent, accuse environmental policy of being a direct antagonist of free enterprise and capitalism. Such an attitude certainly gets in the way of rational discourse and is an impediment to forming cooperative alliances. And actually, the positions of environmentalists mostly encourage freer and more competitive markets and discourage tendencies in the real economy toward central control (as further discussed below).

The association of government control with environmentalism seems to some extent natural, if mistaken. If one believes that environmental protection requires more government control, and if one believes, mistakenly (see Part 2), that more government control means less capitalism and less competition, then an equation of Communism with environmentalism has some elements of logic. And when the environmental community talks repeatedly about profits at the expense of the environment, this doesn't help things.

Yet the environmental record of actual centrally planned economies illustrates the fundamental illogic of an attempt to equate a state-controlled economy and environmentalism. For example, up to its demise the Soviet Union had what was undoubtedly the worst environmental record of any country. From nuclear contamination of vast areas of its countryside to toxic chemicals, from air pollution to the worst nuclear accident the world has known, and to the near loss (and possibly eventual complete loss) of one of the world's largest lakes, the Soviet Union's record regarding environmental protection is indefensible.

Similar patterns are repeated in other countries with greater or lesser degrees of central economic planning. For example, China's air pollution problems are a testament to that government's inability to address environmental issues as effectively as a government presiding over a more capitalist form of enterprise. And to the extent that China is making progress on solving these problems, it is because of the government's efforts to move toward a more market-based economy.

These failures are not accidental: they are a result of the structure of a planned economy within a non-democratic political system.

In a democracy with a market-based economy, industry leaders are subject to market forces and consumer desires, from which they are, by design, isolated in a planned economy. Thus, state planning inevitably will perform worse on environmental issues than capitalism. And an authoritarian government will similarly perform worse than a democratic one. If some conservatives are worried that environmentalism is Communism in green clothing, they can relax.

Social and Psychological Factors

Finally we return to the observation that in a truly competitive economy, it would be impossible to have a unified pro-business and anti-environmental position: that with the proper recognition of self-interest, the business community would almost always be split on the issue of environmental protection, and in some cases would be more or less unanimous in its support. But this is not what we observe.

Why would businesses that actually benefit from environmental protection support policies contrary to their interests through their contributions to trade associations and through their echoing the "party line" on how environmentalism harms business?

The answer is that stronger social and psychological factors than self-interest alone are operating here. Instead of looking at self-interest and acting accordingly, business and the right wing tend to rely on peer pressure and group conformity to set political agendas. Business advocates anti-environmental positions not because they have done the work to show that it hurts their business, but simply because that's what is expected of business people.

Certainly if a business did not have the analytic capability to find and finance two-year pay backs in improving energy efficiency, it would not be surprising if the same failure to engage in strategic planning also led to a failure to identify the right answer to the much more complicated position of how to come down on a public policy issue. Lacking the analytical ability to get the right answer, knee jerking to agreeing with an established industry position that was proposed by a conservative think tank would be easy to understand. Unfortunately, it would not be advancing the corporation's own interests.

If a business lacks the analytic capacity to find win-win options regarding the environment, it will rely on trusted peer groups. And if these peers are influenced by economic fundamentalism that always opposes environment initiatives, its leaders will find it difficult to trust alternative views.

To solve the problems of unleashing environmental protection policy to promote economic growth, trust is an important issue. This book is an attempt, in part, to establish a mutually beneficial basis for improving trust and encouraging collaboration between business and environmental advocates.

The first step to establish trust is to present a concrete case of how environmental policy promotes growth in a large and measurable way. In Chapters 2 and 3 I describe the immense economic benefits of existing and prospective energy-efficiency policies.

2 | Direct Success in Energy Efficiency

Energy efficiency provides the clearest and best-proven example of an environmental-protection policy that enhances economic growth. For energy efficiency, the direct benefits can be seen and measured straightforwardly—they are the value of energy savings. If we invest $1,000 in an energy-efficiency technology that saves $3,000 of energy costs, we have increased our well-being by $2,000, which makes the economy more productive and profitable.

Like any other investment that raises productivity, widespread energy-efficiency improvements the economy undertakes will produce new jobs and increase overall economic growth.

While most people have yet to think of energy efficiency as a large resource, its actual effects are in the range of trillions of dollars. Prior to 1973, energy use was growing in parallel with economic output ("gross domestic product" or GDP). Many analysts predicted that this trend would inevitably persist in the future, and numerous forecasts of future energy needs were based on this premise.

In fact, due to energy-policy activities at the state, regional, and federal levels, and with some small boost from energy price spikes, energy use per unit of economic output began to decrease after 1973. By 2001 it was 42 percent lower than it was at the first energy crisis. About one-half to three-quarters of this decline is attributable to energy-efficiency improvements. Even at the lower end of the range—Vice President Dick Cheney's lower-bound estimate—energy efficiency is saving over 20 percent of America's energy consumption and cost.[1]

In quantitative terms, this amounts to over $125 billion of savings every year. Because the savings will persist for many years, if not decades, the full value of the savings is well over $1 trillion.

Such a result is even more remarkable because it occurred in the face of inconsistent policy attention, or even downright hostility, at the national level. During much of the last thirty years, federal policy did little to facilitate energy-efficiency improvements. The initiative and success began and was sustained at the state level.

Early Resistance to Energy Efficiency

Until the energy crisis of 1973 when an Arab oil embargo sent U.S. oil prices to record highs and created spot shortages that caused Americans to stand in line to buy gasoline, energy efficiency was not an issue of concern to most Americans. Indeed, when a major New York bank performed a significant study in 1972 of the potential of energy efficiency it concluded that the potential for efficiency savings was only about 5 percent because, in general, consumers and producers responded appropriately to the price of energy in their decisions concerning energy usage. In other words, the study asserted almost no opportunities existed to make profitable investments in energy efficiency given the level of energy prices.

However, when America confronted the 1973 energy crisis, and when the issues of energy consumption attracted much more attention, the consequences of evaluating the nature of the crisis, including how to reduce consumption, generated much more sophisticated conclusions. A range of analysts, from respected mainstream organizations like the American Physical Society[2] to such previously unknown researchers as Amory Lovins,[3] concluded that the economy had failed to address energy efficiency as a concern and that the technical potential for efficiency to change outcomes was quite large.

At this point, the very concept of efficiency was poorly defined; the public did not understand the concept well. Saved energy due to a reduction in comfort or performance was not distinguished from technical efficiency, which saved energy because it used enhanced technology that still provided equal comfort or performance.

Policymakers who wanted to dismiss efficiency constantly referred to people's habits of using energy-intensive but fun technologies—big cars and recreational vehicles, long trips, big houses that required excessive heating and cooling—without addressing the issues of efficiency: how many miles per gallon a car achieved if it was more efficient, not smaller; or how to build a house that didn't require climate conditioning to maintain comfort.

In a number of public speeches I gave on the opportunities to reduce energy by applying technology, I used the example of how we could cut refrigerator energy use by more than half, and at the same time improve performance—simply by increasing the effectiveness of key materials: better insulation, efficient motors, improved door-gasket designs, and higher-quality fans.

I repeatedly emphasized that efficiency means providing the same level of energy services—no cutbacks, only better technologies—and gave numerous examples. Yet invariably someone by the end of the speech would ask,, "But what if I don't want to give up the self-defrosting feature to save energy?"

The fact that someone always asked that question shows how foreign even the concept of efficiency was (and still is) to some people. Clearly the questioners were stuck in a worldview where more energy meant better performance, and lengthy discussions of how technology could break that link were unavailing.

If people failed to understand even the concept of efficiency, it would seem unlikely that they would pay for efficient products. So a solution to the energy problems of the 1970s (which are similar to today) would require more active policy intervention than merely providing consumers with information. Many policymakers, particularly at the state level, began to explore which policies were required to address our energy service needs at the least cost to the economy and the least damage to the environment.

Those relatively few agencies and research organizations that bothered to look at energy efficiency discovered a substantial potential: an underestimated resource in the early 1970s—and one that remains underestimated even today at the national and global levels, for many of the same reasons.

We'll describe that resource below, looking first at some specific examples, then at the overall estimates of potential savings (in terms of energy use and dollars), and finally at the state that arguably had tried the hardest and consequently has obtained an impressively high level of realized savings : California.

The Refrigerator Story

The story about how the market approached the issue of energy efficiency, and what the consequences are of policies that encourage energy efficiency, are effectively illustrated in refrigerator energy trends of the past sixty years. Refrigerators are important not only because they reflect how markets work without intervention as well as with different kinds of policy intervention, but also because the refrigerator was the largest single user of household electricity in 1973.

Figures 1 through 3 illustrate that story. Figure 1 below depicts the trend in electricity consumption for refrigerators between the immediate World War II period, when manufacture of these products was reestablished following wartime restrictions, and the mid-1970s.

A History of Increasing Energy Consumption

According to the graph in Figure 1, the energy consumption per refrigerator increased rapidly from about 350 kilowatt hours per year (kWh/yr) in the immediate post-war period to

1,750 kWh/yr by 1972.[4] Energy consumption per refrigerator during this period grew at approximately a 6.5 percent annual rate. When coupled with the growth in ownership of refrigerators (less than 80 percent of families owned a refrigerator immediately following World War II, while the typical family averaged close to 1.2 refrigerators by the mid-1970s) and the increased number of households in the United States, an overall refrigerator energy use growth rate of about 9.5 percent per year resulted.

Perhaps somewhat coincidentally, this was the same as the growth rate of overall residential-sector electricity use. Thus it is reasonable to infer that if anyone had been projecting the energy use of refrigerators as a separate end use in the early 1970s (which no one actually did because forecasts were instead based on aggregate numbers, like total consumption), he or she would have forecasted that an approximate 9.5 percent per year growth rate would continue. And because the growth rates

Figure 1

U.S. Refrigerator Energy Use, 1948–1979

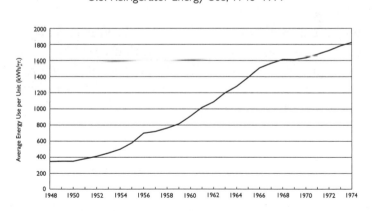

of refrigerator energy use and household electricity use were identical, the forecast would be exactly consistent with government and industry forecasts that were based on extrapolating existing trends. By this calculation, today's refrigerator energy use would exceed 150,000 megawatts of peak-power demand, more than the installed capacity of the entire U.S. nuclear program, a controversial and expensive program that supplies some 20 percent of all the electricity America generates.

Four different factors affected the growth in energy consumption between 1947 and 1972:

- Increase in size. The average refrigerator was about 7 cubic feet in 1947. By 1975 it had grown to about 15 cubic feet.

- Increase in features. In 1947, refrigerators required manual defrosting every few weeks. They had only a single door. Few, if any, additional features were available. By 1975, most refrigerators offered convenient automatic defrosting. This feature increases refrigerator energy use in a number of ways, and can add more than 50 percent to energy use depending on technology.[5] Two-door refrigerators soon offered a separate freezer compartment and replaced the need for a freezer compartment within the refrigerator itself. Options included whether the freezer compartment was on top, on the bottom, or to the side. Additional features included automatic ice and cold water. All such features increase energy use because they provide additional opportunities for heat to enter.[6]

- Increase in performance. A post–World War II refrigerator had a 15-degree Fahrenheit freezer that could not keep ice cream fully frozen. By the 1970s, the overwhelming majority of refrigerators had zero-degree Fahrenheit freezers that could preserve frozen foods nearly indefinitely and provide much higher quality ice-cream storage.

- Decrease in efficiency. This is perhaps the surprising result. A 7-cubic foot refrigerator that had consumed about 350 kWh/yr in 1947 would use twice that amount of energy by 1974. Despite America's optimistic belief that technology always improves performance, the unfortunate result was that improved technologies allowed manufacturers to cut the refrigerator's costs when they degraded its energy performance.

The last factor is particularly interesting. During the 1950s, engineers developed better thermal-insulation materials to allow motors to run hotter (that is, less efficiently) without self-destructing. By testing and deploying features that allowed manufacturers to use less of relatively expensive materials like copper (for the motor), manufacturers reduced refrigerator costs, but at the expense of reduced energy efficiency.

However, the reductions were not cost-effective to the consumer. By the mid-1970s a typical automatic-defrost refrigerator would have cost the consumer roughly $500, but would have consumed well over $2,000 in energy costs over its lifetime. Clearly a 10- or 20-percent reduction in energy cost would have been worth it even if the purchase price increased. Indeed, immediately following the 1973 energy crisis, one manufacturer (Philco) offered a complete line of refrigerators that used 30 to 45 percent less energy than conventional products with a cost increase that paid itself back in energy savings in two to three years. Despite the fact that these refrigerators offered higher profit margins to the retailer, this brand didn't sell well and was withdrawn from the market.

An Energy Consumption Turnaround

The trend toward higher energy consumption reversed rather suddenly in the mid-1970s. (How suddenly is uncertain because the industry did not collect data on energy efficiency or energy performance between 1972 and 1978, and no other reliable data source exists.) However, between 1972 and 1978, energy policymakers were busy addressing the problem of appliance energy use, including but certainly not limited to refrigerators.

The impetus, in part, emerged from states facing critical energy problems, such as California, which was concerned about finding sites for new power plants, and New York, which had experienced the major 1965 blackout (blamed in part on the high growth of electricity demand).[7] The new Gerald Ford Administration in Washington, which was struggling with the issue of how to solve America's energy problems, provided some of the policy direction.

The first tangible result of this discussion occurred in 1975 when the California Energy Commission established mandatory standards for refrigerators—along with central air-conditioners, room air-conditioners, water heaters, and so forth.

State and federal policymakers, as well as private companies and nonprofit organizations, worked together and produced a consensus on how to move forward with energy efficiency. States planned for and enacted mandatory efficiency regulations for major energy users, while the federal government established mandatory energy-performance requirements for some products (automobiles, for example) and set "voluntary" industry targets for improvements for such appliances as refrigerators. (The word voluntary is set in quotes because the law, reflected in the Energy Policy and Conservation Act of 1975, required that there would be consideration of mandatory standards should voluntary targets fail.)

In response to these policy impetuses, refrigerator manufacturers found ways to redesign refrigerators to meet the 1977 and 1979 California standards.

Nationwide refrigerator-efficiency data indicated that manufacturers found it easier to comply with the California regulations throughout the country rather than to establish two different product lines, one for California and one for the other forty-nine states.

Figure 2 illustrates how this experience was just the beginning of a glorious history of innovation.

California's action was the beginning of a long sequence of policy and market interventions designed to improve the energy and environmental performance of the refrigerator. Following the 1980 standards, utilities in California and a few other places began providing consumer rebates for products that went beyond the minimum level the standards required. Figure 2 depicts how the energy use of a refrigerator continued to decline following 1980, albeit at a slower rate—a noteworthy effect the incentives generated.

Figure 2

Impact of Standards on Refrigerator Energy Use, 1947–2002

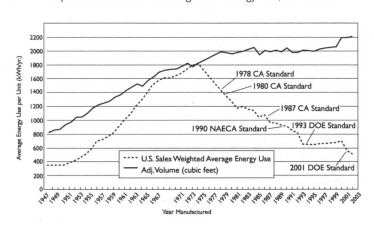

Standards and Incentives for Refrigerator Efficiency: How California Led the Nation

Manufacturers' desire for nationally uniform requirements led to a compromise the U.S. Congress enacted in the National Energy Conservation and Policy Act of 1978 that required the Department of Energy (DOE) to set mandatory efficiency standards for refrigerators and about a dozen other products that previously had been subject only to the voluntary targets. In return, the manufacturers were able to get relief from state standards via preemption.

However, two years later DOE decided not to issue the standards. The Reagan administration DOE opposed standards on philosophical grounds. After trying unsuccessfully to get Congress to repeal the requirement for DOE to set standards, DOE simply refused to set them.

In 1983 the California Energy Commission (CEC), frustrated by the lack of action at the federal level toward mandatory appliance standards, proposed significantly tightening the standards for both the short term (1987) and long term (1992). This regulatory proceeding, which also covered central air conditioners, resulted in a two tiered standard; the first tier with modest requirements for 1987 effectiveness, and the second tier with much more demanding requirements for 1992. The 1987 requirements were based on better-performing refrigerators already available in the marketplace, but the 1992 standards were based solely on engineering studies and, when adopted, no product then on the market in the United States was capable of meeting such standards.

Politically, this aggressive state action drove manufacturers to negotiate with energy-efficiency advocates, including environmental-advocacy organizations, energy-efficiency advocacy organizations, and such state agencies as the California Energy Commission and the New York State Energy Research and Development Authority (NYSERDA), and utilities promoting energy efficiency. These negotiations led to the National Appliance Energy Conservation Act (NAECA) of 1987, which established new standards for 1990 and required DOE to provide periodic additional standard upgrades in the future. The 1993 DOE updates for refrigerators essentially extended the California 1992 standard nationwide.

On an engineering level, manufacturers rose to the challenge of the standards. Some manufacturer representatives had stated to the Energy Commission in 1984 that not only were the proposed 1992 levels not cost effective to the consumer, but also technically impossible to produce. Yet when utility incentives supplemented the 1992 standards, manufacturers both produced products that complied with the standards, as well as creating products 10 and even 15 percent better than the standards. This remarkable demonstration of American industrial ingenuity and know-how came at a net reduction in price (discussed below). Such ingenuity is an impressive tribute to what American industry can accomplish when presented with appropriate incentives.

Ozone Hole Creates a New Environmental Concern: The Need to Solve Two Environmental Problems at Once

In the 1980s, as manufacturers began to think about how to comply with the 1992–93 standards, another environmental challenge arose. During the refrigerator energy-efficiency debate, scientists had discovered that certain chemicals, primarily chlorofluorocarbons (CFCs) that were used as the basis of the refrigeration system and the insulation system of refrigerators, were depleting the stratospheric ozone.

In 1985, the appearance of a giant ozone hole over Antarctica dramatized the seriousness of the environmental problem and prompted governments throughout the world to negotiate a rapid phase-out of ozone-depleting chemicals. This phase-out presented an additional challenge to the refrigerator industry because both the effectiveness of the insulation used in refrigerators and the efficiency of the basic refrigeration system relied on chemicals that would become illegal by the mid-1990s.

Nonprofit organizations, state energy offices, utilities, and the federal Environmental Protection Agency (EPA) were all concerned that an attempt to solve the ozone depletion problem should avoid an exacerbation of the energy problem. Rather than looking at the short-term tradeoffs between efficiency and ozone protection that were the focus of the technical discussions in the 1980s—typical engineering studies predicted a loss of 5 to 10 percent in energy efficiency due to the substitution of more benign chemicals for the soon-to-be-banned ozone depleters—these agencies endeavored to provide manufacturers incentives to consider technologies and design methods that would simultaneously solve both problems. The result was the "Golden Carrot"™ Refrigerator Program: an incentive for manufacturers to design refrigerators that not only meet the 1993 standard but that exceed it by a substantial margin while also phasing out ozone-depleting substances.

The interested organizations formed the Super Efficient Refrigerator Program, Inc. (SERP), a mutual-benefit, nonprofit organization whose goal it was to design the best possible economic incentives for manufacturers to solve the energy and ozone problems.[8] After numerous meetings and discussions with technical experts, the group decided that a winner-take-all competition for a super-efficient refrigerator would be the most effective market mechanism. The amount of money necessary to build a new assembly line to mass-produce super-efficient refrigerators was the basis for the incentive amount.

The contest structure provided that any legitimate manufacturer could bid for a desired incentive per refrigerator and the amount of energy that would be saved: the more the energy savings per unit of incentive payment, the higher the bid score. Complete phase-out of ozone depleters, rather than the 90 to 95 percent phase-out most manufacturers envisioned at the time, earned additional points. Fourteen companies, whose names SERP kept confidential, offered bids, and many of the bidders included the large manufacturers.

The winning manufacturer, Whirlpool Corporation, produced two models of refrigerators, the first saving 30 percent and the second saving 40 percent compared to the 1993 standard.[9] These were delivered over 1994–97. Other manufacturers introduced competing models that were somewhat less efficient, but nevertheless were much better than ordinarily would have been expected given how stringent the 1993 standard was.

This program and its consequences left manufacturers much better informed about the technical potentials for producing consumer-friendly, remarkably high-efficiency products. As a result, for the first time ever, they agreed to negotiate with efficiency advocates on mutually acceptable levels for the next DOE refrigerator standard. The resulting standard cut energy use by 30 percent. Thus refrigerator energy use by 2002 had declined to 520 kWh/yr, less than one-fourth of the level that automatic-defrost refrigerators had used in the mid-1970s when they were smaller.[10] And the most efficient models were 15 or 20 percent better than average. This occurred despite continuing increases in the feature offerings of refrigerators.

The refrigerator story is significant because it illustrates how success in improving energy efficiency or other environmental attributes leads to the potential for further improvement in the future. Such an example contrasts with the more negative paradigm opponents of environmental regulation often cite, wherein a certain limited number of measures exist, and once exhausted, no further progress is possible. The refrigerator story leads to a paradigm for energy efficiency that is much more like the model of labor productivity improvements, in which one round of implementation of efficiency helps management envision a next round.

Many efficiency advocates refer to policies of acquiring cost-effective efficiency measures before proceeding to more expensive new energy supplies as "picking the low-hanging fruit." The refrigerator story shows that the metaphorical efficiency tree is even more bountiful than people had thought: we find that after picking all the low-hanging fruit, it grows back. Thus an analysis of the opportunities for energy savings from all appliances in 2004 found that another round of refrigerator efficiency standards was still one of the most attractive options available in the United States.

The refrigerator story deserves prominence not only in that it represents one of the biggest success stories of energy efficiency over the years, but also because nothing unique exists about refrigerators; there is every reason to anticipate that the lessons from refrigerators would be applicable to virtually any other user of energy. We have made more progress with refrigerators simply because, with insufficient national attention on energy efficiency, policymakers had to first focus on the highest priority savings, at the expense of downplaying, or even ignoring, lower-priority, though still important, opportunities.

Other Energy-Efficiency Opportunities

The United States has made similar progress in a number of other areas. The California Energy Commission prepared an illustration to depict the progress toward efficiency of two other major products (Figure 3). We again see the pattern that standards lead to major improvements in energy efficiency, and in some cases incentives provide modest yet still significant improvements.

Air Conditioning and Heating

For some end use, such as air conditioning and heating, even greater progress can be made than the refrigerator story illustrates, because we are not dependent on just one component of the system—the air conditioner or the heater—but rather we can achieve energy savings from the larger system of cooling and heating a home. To reduce the need for air conditioning in the first place, improve the insulation and air tightness of the house, employ windows that reflect the heat of the sun without looking dark or mirror-like, and equip the home with more efficient lighting and appliances that produce less heat These measures lead to overall savings that reflect both the r.educed need for the air conditioner and the reduced energy use for each unit of air conditioning.

Figure 3
Effects of Standards on Efficiency of Three Appliances

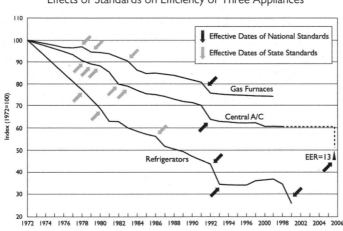

Source: California Energy Commission

Figure 4 shows the improvement in energy efficiency for cooling a house in California. It shows both the effect of the air-conditioner standards adopted at state and federal levels and the effect of periodic upgrades in the state's building energy-efficiency code for new homes. What is more remarkable is that these reductions were achieved in the face of a substantial increase in the size of a new home over the period covered in the graph.

Figure 4
Annual Energy Use of Air Conditioning in New Homes in Calfiornia

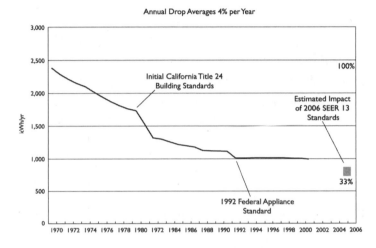

Source: CBC Demand Analysis Office

Efficient Lighting in Commercial Buildings

More than 100,000 megawatts of electric power—well over 10 percent of all peak power use in America—comes from lighting commercial buildings in the daytime. The opportunities for improving the efficiency of this sector are immense. Utilities that operate energy-efficiency programs generally find that a disproportionate share of the savings from their whole program comes from commercial lighting, and that the value of energy savings per dollar spent on these programs is the highest.

In the 1970s it was not uncommon for office buildings to use 3, 4, or even 5 watts of lighting power for each square foot of space. Lights would run twenty-four hours a day, even when the workers were only at their jobs from 9 to 5. (I can look out from my living room window at the downtown San Francisco high-rises, and in the late 1970s nearly all the lights in the windows would still be blazing away at 1 a.m.)

In many cases, neither laziness nor an unwillingness to turn off the lights caused 24-hour-a-day operation. It was because the electrical system did not allow workers access to light switches. Sometimes a whole floor of a large building would have only one switch, or unmarked switches were located in a remote area so you couldn't tell which switch affected which office. And sometimes the switches were locked behind secured boxes or doors.

By the 1980s, energy codes promoted lighting designs that used only 1.5 to 2 watts per square foot and required accessible switches. While this was a great improvement, it was not anywhere near where technology allowed us to go. At this time I proposed that a cost-effective target for lighting power was 0.5 watts per square foot, and published this recommendation in a construction industry magazine.[11]

The lighting industry was shocked at my recommendation and was nearly unanimous in its insistence that it was impossible. But in 1987 we had the opportunity to demonstrate the 0.5-watt-per-square-foot office during a gut-renovation that the Natural Resources Defense Council (NRDC), an organization I was working for, was performing. We employed a noted lighting designer who, after some initial skepticism, agreed to do the work after reviewing why we thought the plan would succeed.

The building did meet the goal, and after construction, even one of the main doubters had to admit after touring the building that the quality of lighting was good and that the designers and owners had met their energy goal.

How did we do it? We avoided the lens-covered fixtures that were the dominant fixture design of the 1980s, but which retain the heat of the four lamps they contain, making them less efficient. Such fixtures also waste light because the lens absorbs it, and they tend to create glare. Plus the lights are too bright for comfort.

We used the new generation of thinner 1-inch lamps that allow better control of light, produce more light per watt, and also provide better color rendition. We replaced metal-wound transformer ballasts with electronic ballasts, which eliminate flicker as well as cut energy use, and provided controls that shut off the lights whenever an office is unoccupied. We used fixtures that employ reflectors around the lamps for low-ceiling rooms and suspended fixtures that reflect some of the light off the (white) ceiling for high-ceiling rooms. And we employed task lighting, where only the tasks are illuminated at high levels while the background is at lower, comfortable lighting levels. Task lighting has been the recommended practice of lighting design guidelines in North America and in Europe for decades, but is seldom used in the United States.

Such lighting techniques are still not commonly used in 2005 even though they represent attractive efficiency opportunities. However, the newer generation of fixtures, lamps, and ballasts is significantly better both in energy efficiency and in attractiveness than those 1980s state-of-the-art items, so even more savings potential exists now.

More Energy-Efficient Automobiles

The transportation sector offers even more diverse opportunities to improve energy efficiency and the quality of life. The transportation sector accounts for some 25 percent of U.S. energy use. Private transportation, through cars, light trucks, SUVs, et cetera, is the lion's share of this and accounts for some 18 percent of energy use. (Note: despite the high degree of public attention on personal transportation, buildings use approximately twice the energy as automobiles.)

Energy efficiency in the transportation sector can be increased in two distinct ways. The first is to produce more efficient automobiles. The U.S. Congress required that the auto manufacturers double fuel economy between 1975 and 1985. Despite predictions of economic ruin that the auto industry asserted, these changes were accomplished over the next ten years, largely without reduction in vehicle size or performance. Most likely as a result, gasoline prices and world oil prices declined. (If demand for gasoline drops due to higher fuel economy, the ability of producers and of OPEC to command higher prices is reduced, and prices will have to settle at a lower level. Conversely, in 2005 rising oil consumption due to China's fast economic development, as well as the stagnation of the U.S. automobile fuel economy, emerged as likely causes for high oil prices.)

Energy-efficiency opportunities in cars are similar to those in other technologies: market failures prevent efficiency improvements with returns on investment of 30 and 50 percent from selling well. In fact, it is so widely understood among the auto manufacturers and their dealers that efficiency fails to enhance sales, that in most cases options that improve efficiency and raise vehicle costs slightly (so slightly that the extra costs are paid back in gasoline savings in two to three years) are unavailable.

Several studies have documented how conventional and well-known technologies can raise typical fuel economy from about 25 to 40 miles per gallon while rapidly paying back their additional costs to the consumer. In evaluating such studies, recall that similar reports for refrigerators, air conditioners, clothes washers, and other products where the efficiency recommendations were actually implemented in policy, projected costs that were much higher than the actual increase in cost.

Hybrid Cars

Beyond the use of conventional technologies, recent products have demonstrated the viability of so-called hybrid—gas/electric drives—featuring a gasoline engine that also charges a battery to run the car whenever its power needs are low enough.

The technical reason why hybrid drive works is that, while automobile engines are very efficient and clean at a particular combination of low engine speed (rpm) and high torque, the efficiency drops off (and the engine also produces much higher air-pollution emissions) under the low-load, low-torque conditions typical of most driving. Thus a car idling at a red light consumes almost as much gas per hour as the same car driving at 30-miles-an-hour constant speed.

Hybrid-drive technology encourages the use of the gasoline engine only at its most efficient operating point to solve this energy-efficiency problem. In an ideal hybrid, the engine would only operate when it was near its most efficient conditions of torque and rpm. Electric power from batteries would be used at all other times. As of 2005, real hybrids do not achieve this ideal performance, but the most advanced do allow the engine to turn off when the electric motor is capable of handling the need for acceleration (such as starting from a complete stop). For the most advanced hybrid drive available as of mid-2005, this technology appears to improve fuel economy by almost 50 percent.[12]

Hybrid drive allows manufacturers to obtain further savings from technologies that fail to improve fuel economy significantly with conventional engines and drive trains. For example, if a car always operates at an inefficient point of its engine's performance map, improving the efficiency of the car—by decreasing its rolling resistance through better tire design, for example, or by improving its aerodynamics at low speed—offers little benefit. Sure, they reduce the load on the engine, but because the efficiency of the engine goes down as the load goes down, the savings in gasoline is insubstantial.

Yet with a hybrid drive these technologies start to make sense. The technology opens the door to other opportunities for energy-efficiency improvement that wouldn't make technical sense with conventional engines and drives. One interesting example of such technology is improved air conditioning systems.

Automotive Air Conditioning

There is no motivation for manufacturers to make automotive air conditioners very efficient at present. This is the case for both policy and technical reasons. The policy reason is that the EPA miles-per-gallon test is performed with the air conditioner off. Even if more efficient air conditioning saved lots of gas, no one would know, because the savings are not accounted for in the fuel-economy rating. Manufacturers therefore cannot use efficient air conditioners to meet the fuel-economy standards, and consumers have no ability to comparison shop between models with better or worse air-conditioner efficiency.

The technical reason auto air conditioners remain inefficient is because, again, the additional load imposed on the engine due to air-conditioner operation also increases the efficiency of the engine in most circumstances, so the gasoline savings for more efficient air conditioners is small.

How would you redesign the air conditioning system of a car if a 50-percent reduction in energy needs actually saved 50 percent of the gas? The first thing you would notice, which anyone who has ever parked a car on a hot, sunny day knows intuitively, is that a car is basically a solar oven. On a hot summer day, sunshine streams through the windows, is trapped inside (particularly if the interior is a dark color), and interior temperatures can exceed 160 degrees Fahrenheit. The increased temperature is not only uncomfortable when occupants return to the car, but also deadly for small children and pets inadvertently left alone in such a parked car.

A sensibly designed automobile air conditioning system would provide solar-reflective windows that keep out half of the solar heat without reduced visibility. With less heat entering the car, the opaque parts of the auto body could be insulated better to minimize heat gains from the hot outside air.[13]

Advanced technologies could make windows nearly 100-percent reflective when the car is parked, yet fully transparent as the car drives. Perhaps an intermediate position that provided the driver with virtual sunglasses would both enhance safety and reduce air conditioning load while driving.

Finally, with greatly reduced heat gain from the sun, manufacturers could put a small solar cell in the roof of the car and use a solar-powered ventilation fan to cool the car when it is parked with the motor off.

Researchers at Lawrence Berkeley National Laboratory have estimated that the use of such technologies would lower the inside temperature of a car parked all day in a hot parking lot to only a few degrees warmer than the outside temperature. In addition to reducing safety hazards, driving on hot days could be far more comfortable, too.

These technologies would save significant amounts of gasoline for air conditioning in a hybrid, but not in a conventional car. In a conventional car, the reduction in air conditioning load on the motor would almost always reduce its efficiency by so much that the consumption of gasoline would scarcely be changed. Clearly the incorporation of a new efficiency feature into a product—in this case hybrid drive into a car—often does not exhaust the efficiency potential, but instead makes possible the addition of other efficiencies.

Location Efficiency

If we enhance the "location efficiency" of our urban and suburban neighborhoods, large additional savings in gasoline use are possible. Americans drive about twice as much as Europeans, and about four times the amount the Japanese drive. Typical justifications include the wide-open spaces in America and low gasoline prices, but neither of these arguments holds up to scrutiny. Instead, different forms of urban development best explain the differences: other countries make greater use of neighborhood and transportation system designs that minimize the need to drive.

Recent studies have also documented ratios as high as 5:1 between how much Americans drive in different neighborhoods within the same city: some neighborhoods allow people to reduce their driving by 80 percent compared to comparable families (same size, same income) living in suburban sprawl. A number of analyses of driving habits throughout the world have linked what is now called smart growth development patterns with reductions in the need to drive. Smart growth emphasizes compact development with walkable and bikable streets and the provision of high levels of mass-transit service.

One particularly interesting study in which the author participated examined the differences in driving behavior among different neighborhoods in three large metropolitan areas in the United States: San Francisco, Los Angeles, and Chicago.[14] The studies covered the outermost suburbs, as well as the inner city, and focused on every single neighborhood within the urban area. The measures of driving were the number of vehicles owned per household (a U.S. Census measure, which documented the number of available cars in representative households) and the number of miles driven per car, which was available—on a confidential basis—via smog-check records of odometer readings from cars that are required to be certified every two years. Over six million records of odometer readings were examined.

The study attempted to draw statistical connections between the number of vehicles owned and the number of miles driven with different variables that could explain them. Included among the variables were household income and household size along with such smart-growth variables as the compactness of development (number of housing units per acre of residential land), the level of transit-service availability (number of buses or railcars per hour that stopped within walking distance of the house over the course of a day), the pedestrian and bicycle friendliness, the availability of local neighborhood jobs, and the proximity to metropolitan area jobs (number of jobs available within a half-hour drive). The study used statistical methods to predict car ownership and miles traveled as a function of these variables.

It turned out that four variables were extremely significant statistically, two variables had modest statistical significance, and the rest of the variables had no definitive effect. The four most significant variables were:

1. Compactness of development
2. Transit service levels
3. Income

4. Household size

The other significant variables were:

5. Pedestrian/bicycle friendliness

6. Proximity to jobs

The most important variable explaining driving behavior was the compactness of development. The higher the number of housing units per acre, the less people needed to drive. Over reasonable ranges of compactness measured in the study, the need to drive could be reduced by almost two-thirds compared to the extreme case of urban sprawl (about 2 to 3 units per acre).

Three other variables were almost as important and of about equal statistical significance: transit service level (households within walking distance of a metro stop needed to drive about 30 percent less than those lacking transit service, regardless of the neighborhood density), income (the propensity to drive increased strongly with higher annual income up to about $100,000, after which income had no effect), and household size (not surprisingly, the more people in the household, the more driving was necessary, with major differences in household sizes ranging from 1 to 3 or 4).

The two smart-growth variables above had remarkably strong influence on reducing the number of cars people owned; they also reduced the number of miles driven per car, although the range of variability was much smaller.

Two other variables—pedestrian/bicycle friendliness and proximity to jobs—were also statistically significant, but with lower confidence than the previous two smart-growth variables. Neither had an effect on car ownership, nor did they reduce the number of miles driven by more than about 5 percent.

One particularly noteworthy observation based on these results is that expanding transit service levels, which is possible to do relatively quickly (simply put more buses on the street), reduced driving disproportionately more than expected. We note that locating a metro stop within walking distance of a house reduces the need to drive by 30 percent. However, far fewer than 30 percent of the families within walking distance of a transit stop ever use the metro, much less rely on it for all of their transportation. Apparently the existence of good transit service changes the driving behaviors of people who never use the transit system, which reduces the need to drive and improves accessibility for all.

While a statistical study cannot explain how or why this happens, one intuitive answer is that if good transit exists, retail services tend to concentrate near the transit stops, so a driver in a transit-rich area can make one trip to a retail area, perform several tasks, and drive home. In contrast, in an area with poor transit, there is no reason to concentrate retail service, so people drive from one store to another to another. Also, with good transit, a resident is not stranded if no car is available at a given time, so families are able to meet their transportation needs with two cars rather than three (or three cars rather than four).

Smart Growth

These results are not surprising to anyone who has traveled throughout the world and observed cities that have different levels of smartness of growth. What is surprising is the extent of explanatory power in the statistics. First of all, the equations that predict car ownership and miles driven were almost identical among all of the three cities. The same patterns persisted in the colder climate (Chicago) compared to the milder climates. Additional supporting studies also showed that the effects were similar regardless of stage of life (whether a family is married with children, single, young, or retired), and similar patterns appear to occur in other regions of the United States and throughout the world. And the level of correlation between smart growth and reductions in car ownership is amazingly high compared to the results of other studies of human behavior.

The overall significance in terms of energy and the economy is immense. A recent study showed that if all new suburban and inner-city development in the United States for the next ten years were smart growth, U.S. gasoline consumption overall could be cut by 8 percent.[15] Savings would steadily grow larger over the years as smart growth development increased because neighborhoods last hundred years or longer.

The potential value of transportation expenses saved due to smart growth can be impressive. Surveys of American-family household expenditures indicate transportation is the second-largest expenditure. At some 18 percent of income, transportation is second only to housing costs. In most cities, the combination of housing and transportation expenses accounts for about 50 percent of overall household expenses, slightly higher in cities most affected by urban sprawl and slightly lower in cities with smart growth urban corridors. The typical cost of transportation for a household in suburban sprawl exceeds $8,000/yr even when gasoline prices are well below $2 a gallon.

(This $8,000/yr number is remarkable. It means that when a typical family purchases a median-price, new home in suburban sprawl it will spend more getting to and going from the home over the 30-year mortgage than the home originally cost[16]!)

In contrast, smart growth development patterns can reduce such cost by 30 to 75 percent. Additionally, the reduction creates enormous improvements in purchasing power for the family. It also enhances local economic development because much of the avoided costs go outside the region, and often outside of the United States (what with foreign sources supplying all of the incremental energy and an increasing share of the automobiles and car parts).

If all new development over the next ten years were smart growth the overall transportation savings to the American economy would exceed $2 trillion.

Efficiency in Industry

Industrial energy use is more difficult to characterize than residential or commercial energy use because most industrial energy uses are specific to a particular plant. Yet the same magnitude of opportunity exists in process industries as in the other sectors.

Most industrial electricity use is to run motors, which in turn operate such mechanical systems as drives or compressed-air systems. Each of these drives is different; that is, they run a different type of process. Some may operate machine tools, while others run compressors for refrigeration systems

There are a number of ways to save motor energy. The easiest way to understand is to buy a more efficient motor. The next is to look at how the system performs when the motor can operate at below maximum power. Many motors are poorly controlled, so when they are not needed at full power they turn off and on. Home air conditioners are an example, as are the air conditioners that operate in small areas of industrial plants. Unless motors are designed with control devices that allow them to run more slowly, their performance will be inefficient when not operating at full capacity.

Often there are major savings potentials in the drive system. Many compressed-air systems are sized incorrectly or leak. And in some cases the very concept of compressed air may be obsolete compared to using smaller motors at each machine or using a mechanical drive. And many mechanical drives are also obsolete, losing energy by slippage. For example, leather or rubber drives may slip on the wheels they are driving, wasting energy compared to employing a separate electric motor on each wheel.

The largest overall use of industry energy is for heating. Heat is needed primarily for such industrial chemical processes as steel production, petroleum refining, glass melting, and a wide variety of chemical needs.

Simple ways to save heat energy include the use of higher efficiency boilers or furnaces, reductions in leakage from steam-distribution systems, and better insulated pipes.

Industrial facilities are also major users of lighting. The same principles that guide efficient design of lighting systems in commercial buildings also apply for industrial facilities, although the appropriate choices of technologies are different. Yet because the principles are the same, opportunities to save 30 to 50 percent or more of industrial lighting energy at a return on investment of 50 percent annually are widespread.

A more fundamental way to save industrial energy is to look at the entire process, as discussed in Chapter 3: "How Business Ignores Process Engineering."

This discussion of industrial energy use will not get into the details because the most easily understood efficiency measures, such as insulation and better lighting, are similar to those available for homes or commercial buildings, while the most effective are each unique. Interestingly the discussions of energy-efficiency opportunities in industry often overlap with discussions of other ways to improve industrial environmental performance (by reducing pollution emissions, for example) in profitable ways. Such discussions also support this chapter's main point: that many unexploited opportunities to make profitable investments in industrial plants can be found that improve the environment, and that even more opportunities will emerge if environmental policy spurs innovation.[17]

Many companies already pursue these potentials, but they remain in the minority.[18] Even some of the best-managed companies, even those that achieve great gains in efficiency in one area, miss the boat in others.

How Far Can We Go with Efficiency?

We now have overwhelming evidence that demonstrates a large and growing potential for energy efficiency to reduce economic costs and the emissions of pollution associated with energy production. Efficiency can reduce energy use by 75 percent over thirty years even when the product that uses energy grows and the level of energy services increases. Such savings are possible on a regional or national scale and are remarkably close to the predicted level.

We have also seen strong evidence that these efficiency measures and policies reduce overall consumer costs—both to the business consumer and to the householder—and to society overall. The cost reductions are large—in the tens of billions of dollars annually for a state and in the hundreds of billions of dollars per year for the U.S. as a whole. Clearly these cost savings produce growth in output and in jobs because if we receive the same energy services at lower cost the difference can be spent on new economic activities, whether they are consumption or investment.

Next we explore how big these efficiency resources are and determine how realistic it is to expect policies to get us there, first in theoretical terms (on paper), then in actual terms (in reality). The studies on paper are prospective—they look at what can be accomplished in the future. The studies that examine reality necessarily are retrospective. However, we can make comparisons on an apples-to-apples basis because the paper studies have been done since the 1970s.

How Far Can We Go on Paper?

It's a simple and straightforward matter to add all of the energy-efficiency opportunities for all the major uses of energy—there aren't more than fifty or a hundred important end uses to analyze—and theoretically project what energy use for the whole economy might be if more energy efficiency were obtained through policy.

Perhaps the most interesting of these studies is America's Energy Choices, a joint effort of four environmental advocacy organizations in the early 1990s. This report is significant for the level of technical detail that it documented. The technical appendices show the specific technologies that are applied to a large number of end uses and what the costs and benefits of each technology are. This allows for more careful review and evaluation of the study than is typical. It also allows detailed calculations of the economy-wide results.

The results are summarized in Figure 5: after forty years, the environmentally preferred scenario would reduce U.S. emissions of global-warming pollutants by 70 percent.

Figure 5
CO2 Emissions Under America's Energy Choices' Scenarios

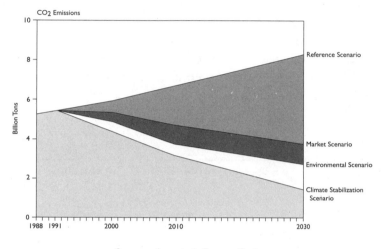

Source: *America's Energy Choices*

Emissions of health-related air pollutions would decline even more. We might expect that the costs of these emissions reductions would increase as the scenario becomes "greener," but actually the reverse is true. Figure 6 displays the cost of efficiency and renewable energy measures to the economy over the next forty years. While the investments required grow from $1.3 trillion to $2.7 trillion as the emissions reduction level is increased, the savings grow even faster. The net economic benefit of this most aggressive scenario—$2.3 trillion—exceed the net economic benefit of the other two less-aggressive scenarios, so despite the inclusion of some measures that are less cost effective, the most environmentally beneficial scenario is the most economically attractive.

Further analysis of this scenario showed that the most environmentally beneficial scenario would also create one million more jobs than business-as-usual.

These economic conclusions are remarkable because they do

not take account of the strong form of the economic growth proposition, which will be discussed in depth in the next chapter. The report did not tabulate the nonenergy benefits of efficiency measures. It did not take credit for the fact that the realized costs of efficiency usually are less than the projected costs. And it did not analyze how much energy prices would drop because of reduced demand. Despite these conservative assumptions, the magnitude of economic benefits is still large.

Figure 6
Costs and Saving Scenarios

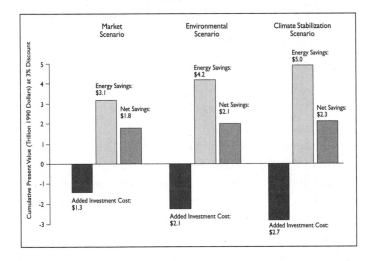

Source: *America's Energy Choices*

Other organizations and agencies have also performed analyses of how far we can go on paper, and reach similar conclusions. The California Energy Commission and the Northwest Power Planning Council performed perhaps the most significant of these studies because they formed the basis for actual policy decisions, such as how much money to spend on efficiency and how many power plants are needed. For over two decades, these studies consistently show that efficiency is the cheapest and usually the largest source of new energy.[19]

How Far Can We Go in the Real World?

It is one thing to have a number of studies demonstrating that something is possible on paper; but it is another thing to actually implement the results in the real world. How do we know that the optimistic results of energy studies are achievable? Perhaps the best real-world demonstration that the studies are on target is California.

Recall that California was perhaps the first state to initiate comprehensive studies of the potential for energy efficiency (and renewable energy) to ease the burden of constructing too many conventional power plants and other energy supplies. Beginning in the early 1970s California, with bipartisan support, established new policies at its Public Utilities Commission (CPUC) and created a new agency, the California Energy Commission (CEC), to promote energy efficiency.

One of the first activities of the new California Energy Commission was to develop the data and methods needed to do an end-use-based forecast that could then lead to informed decision-making about both the need for new energy-supply facilities and the ability of energy efficiency and renewables policies to change that answer. The Commission also immediately initiated actions that led to the establishment of efficiency standards for new construction and new appliances and equipment. At the same time, the California Public Utilities Commission made changes in the structure of utility regulation that encouraged utilities to look more thoroughly at efficiency options, as well as conventional supply-side options as a means of meeting their obligation to serve at the lowest cost.

Figures 7 and 8 illustrate the results of these policy changes. Figure 7 shows the effect of California's new energy policies in terms of energy savings.[20] Both regulation-based and incentive-based policies prove to be remarkably effective, and the amount of power they have saved amounts to

15 percent of California's electricity use. Other policies that encourage the most efficient and cost-effective fuel for each end use further restricted the growth of electricity use. Figure 8 shows that California has held its per-capita consumption of electricity flat for the last thirty years while in the rest of the country, electricity consumption grew by 50 percent.

This is a remarkable difference, for two reasons. First, the United States as a whole has made significant progress toward realizing the potential savings from energy efficiency, even if the federal government wasn't trying its best.[21] According to the Bush Administration's National Energy Plan of 2001, energy efficiency was the largest new source of energy since 1975, accounting for at least 20 percent of total usage, as noted previously.

Second, the rest of the United States has some other shining stars of efficiency—other states have records of achievement as good as or almost as good as that of California—that move up the national performance average.

So California's pulling ahead of the U.S. average as shown in Figure 8 is all the more remarkable in light of the fact that the rest of the U.S. was not standing still. It is also impressive because many of the most effective California policies—such as the 1992 refrigerator efficiency standards—were nationalized soon after their adoption in California. So the state had to run faster and faster to stay ahead. But Figure 8 shows that it has done so.

So the fact that California is consistently beating the nation as in whole in the efficiency race shows just how far a jurisdiction can go if it tries, even given political realities.

These dramatic results are consistent with relatively aggressive forecasts of potential energy efficiency that have been made throughout the last thirty years. In sum, this experience shows irrefutably that on a scale as large as a nation savings on paper can be translated into savings at the meter. (California would be about the sixth-largest economy in the world if considered as a separate nation.)

Figure 7
Energy Savings from California's Energy Policies

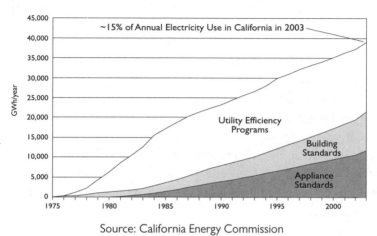

Source: California Energy Commission

Figure 8
Per Capita Electricity Sales in California and the U.S.

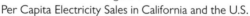
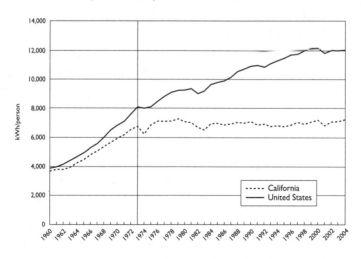

Source: California Energy Commission

It is particularly noteworthy that this dramatic record of success in recognizing and using energy efficiency as an economic development strategy has been achieved with bipartisan support and with a much lower level of controversy than at the national level. Figures 7 and 8 show continual progress in improving energy efficiency. We see neither accelerations in efficiency under one political party's leadership nor slowdowns under another's: both graphs illustrate continued progress regardless of which party is in control. Regardless of which party's appointees ran the California Energy Commission, its planning documents consistently identified energy efficiency as a good energy-planning goal, and, more than that, as a means for promoting economic development and growth in the state.

The California story is even more interesting in that most of the data is recorded prior to 2001. Both Chapter 5 and Appendix A address how California faced the prospect of a critical electricity shortage in 2000. In response, the state initiated an emergency program of efficiency and conservation that entirely averted the problem. It quadrupled the budget for incentive-based energy efficiency as the cornerstone of its plan to accomplish this victory.

Evaluations of the results showed that a four-fold increase in the efficiency budget produced a four-fold increase in savings. Efficiency program managers noted that even with four times the previous years' budgets, they ran out of money before they ran out of opportunity.

An experiment by PacifiCorp, a large utility serving the Northwest, performed in the early 1980s offers a similar result, on a smaller scale. The object of the experiment was to see how far we could go with home-heating retrofits if there were no longer any barriers to participation. PacifiCorp, the Bonneville Power Administration, and public-interested participants designed a package of advanced insulation measures for households, and offered to install the package for free in the Hood River, Oregon, community.

After the three-year program, over 85 percent of the homes had been retrofitted, and energy use for heating was dramatically lower than in previous years[22]. Of course, extending this program to such a broad area as a state or a nation would be expensive in terms of the amount of incentive budget needed, but the experiment proved that the energy savings could be achieved quickly in the real world and could be obtained without any undesirable side effects.

Such energy-efficiency examples confirm clearly and firmly that environmental protection leads directly to enhanced economic growth and jobs. Both in the studies that are based on calculations and on engineering and in the real-world experience illustrated so strikingly in California, we see that efficiency policies save large amounts of money, which will be used to create growth and new jobs rather than wasted on uneconomic sources of energy supply.

The story is important because even in California neither state government nor the business community (or even the environmental community) have ever given energy efficiency top priority. While there have been a number of successful programs over the years, none of them has ever achieved the full amount of efficiency that is cost-effective and practical.

The limits to how far a state or a nation can go with energy efficiency have never been tested in the real world.

The few tests that have been run are always limited by budgets, rather than by technology or consumer or producer response.

This chapter addresses the opportunities to improve energy efficiency throughout the American economy. Similar stories apply throughout the world, particularly because as countries move from impoverished status to more economically successful conditions, they rely increasingly on internationally similar or even identical (as in the case of lighting) products for which identical efficiency methods exist.

The next chapter presents how the opportunities have usually proven to be even greater. Unexpected benefits have emerged from energy-efficiency policies that have been carried out and evaluated, side effects of the policy that both increase the benefits and reduce the costs, often to zero. This observation suggests an even larger role for environmental policies in growing the economy and producing jobs.

3 | Enhanced Innovation— Energy Efficiency's Unexpected Success

In addition to the direct energy-saving benefits outlined in the previous chapter, significant and even more substantial indirect economic gains are possible from policies that promote energy efficiency in particular and environmental protection more broadly.

Chapter 2 identifies several of the countless opportunities to invest \$1 in energy efficiency and save \$2 or \$3 in energy costs over the life of the building or device. This chapter presents how in many cases the value of nonenergy benefits that were not considered previously overshadow the \$2 or \$3 in energy savings—and the predicted \$1 cost increase doesn't come to pass.

Such nonenergy benefits come from the ability of environmental policies to promote innovation and competition. So when we try to introduce known technologies that provide known environmental or energy benefits, such as refrigerators whose operating costs are one-fourth of what they were in the 1970s, we also produce side effects that are often even bigger and better than the direct effects.

The first of such side effects is nonenergy benefits: advantages to the consumer and to the bottom line of the companies that use more energy efficient products.

A second side effect is the promotion of innovation in industrial processes. As policies encourage or require industries, companies, or individuals to change their behaviors to look for better technologies, they start looking for other behaviors that can change at the same time. And they usually discover many. So, for example, if a window manufacturer has to change its assembly line to replace aluminum framing with vinyl or wood for energy efficiency, it can simultaneously consider new ways to reduce materials cost, improve worker health and productivity, and change what parts it buys as opposed to what it makes in-house.

Inquiries into how to make processes more productive are called process engineering. Process engineering seldom occurs in America, so to encourage more is another important pathway for environmental policies to encourage economic growth.

More broadly, many of the world's most successful economies over the past fifty years seem to have profited from increasing their resource efficiency—substituting creativity for (imported) resources. This chapter looks at this substitution as another mechanism for enhancing growth.

Nonenergy Benefits

Why are nonenergy benefits so important to the discussion about how environmental policies promote economic growth? The reason is because they allow the size of the boost in growth to be much larger than it would be otherwise. Energy costs are about 7 percent of the economy. So if policy can reduce these costs, the amount of additional economic growth that will be stimulated is important but not necessarily overwhelming.

However, if energy efficiency also regularly produces nonenergy benefits, the size of the economic stimulus from the promotion of these benefits can increase greatly. Over the past forty years we have found most commonly that as newly incorporated technologies and designs improve the energy efficiency of products, nonenergy benefits come along for free with the energy benefits. In many cases, technical reasons suggest why this ought to be the true: but in other cases, no obvious inherent reasons exist why higher energy-efficiency options ought to perform better—but they usually do.

Perhaps it is the process of innovation triggered by the need to meet energy or environmental criteria that inspires manufacturers to find ways to improve other consumer features of the product. Or perhaps there are other aspects of efficiency that make efficient products perform better overall.

Whatever the reason for the prevalence of nonenergy benefits, organizations that promote the efficient products do so not ordinarily on the basis of energy savings or environmental benefit, but rather on the basis of consumer features that are indirectly energy-related. Such a focus on nonenergy benefits is common to all interested parties, including their manufacturers and retailers, utilities that promote energy efficiency, and state energy agencies that promote energy-efficient products.

Here are some examples of the kind of innovation that has produced nonenergy benefits:

1. Efficient clothes washers. Since 1992 the new generation of efficient clothes washers has been marketed based primarily on the washers' stain-removal ability or size and capacity options. Energy-efficient washers don't require an agitator and therefore can accommodate more and bulkier garments. In addition, they produce water savings that often reduce consumer expenses even more than the energy savings. And because they use less water, they need less detergent; so detergent savings may be larger still. Finally the gentler cycle of washers without an agitator allows not only longer life for all our clothing, but also permits home washing of some garments that previously required dry cleaning.

2. Energy-efficient dishwashers are marketed based on how quietly they operate. They also require less water and less detergent.

3. Compact fluorescent lamps are promoted with the claim that they provide the same quality and quantity of light as incandescent ones but offer longer bulb life. Compact fluorescents typically last six to ten times as long as standard light bulbs. Some are marketed with five-year warranties.

Because of the importance of nonenergy features, the Energy Star® program for compact fluorescent lamps bases its specification primarily on such performance indicators as color, quality, ability of the lamp to turn on instantly, lifetime, absence of flicker and noise, et cetera, rather than primarily on energy efficiency.

The reason for this choice of specification is that in order for compact fluorescent lamps to be acceptable in the marketplace, they have to provide strong consumer satisfaction, and these specifications assure that a first-time consumer will likely be happy and buy more in the future. Energy efficiency is less of an issue because even the worst compact fluorescent lamps save two-thirds of the energy compared to an incandescent.

Notice also the labeling and packaging of compact fluorescents: they are sold primarily on the basis of consumer benefits, such as longevity.

Interestingly, for a commercial user of compact fluorescents, the primary direct benefit indeed is long life because the bulbs require fewer replacements. The indirect savings in maintenance staff time required to change light bulbs is even higher than the direct value of energy savings.

The lifespan of compact fluorescents at their best can be astonishing. When my children were young, we installed compact fluorescent track lights in the corridor outside their bedroom, an area that is illuminated almost every evening. Eighteen years later, we've changed only one of the two bulbs.

I also discovered an unexpected benefit of compact fluorescent lamps when I noted to my friend, who's a physicist, that my glass lighting fixtures were dirtier when I used ordinary light bulbs. He explained that this is a consequence of a well-understood phenomenon, called thermophoresis. The molecules on hot surfaces move faster than those on cooler surfaces, so a hot surface will cause air molecules to move faster when they bounce off the surface; thus, dust particles are pushed away from a hot surface. The reverse situation occurs at a cool surface: molecules move more slowly and air molecules will push dust toward the cool surface.

In a light fixture, the hot surface is the light bulb and the cool surface is the fixture. So a more efficient bulb generates less heat and has a cooler surface temperature; the bulb does not push dust to plate out on the fixture nearly as much. So compact fluorescent fixtures stay clean much longer than ones that use old-fashioned light bulbs.

The story of nonenergy benefits is just one part of a centuries-old pattern of improved lighting efficiency and nonenergy benefits. In the eighteenth century, candles and oil lamps were the predominant lighting sources. When Thomas Edison invented the light bulb, its efficiency was almost twenty times higher than candles. But it also was much less of a fire hazard. Candles are expensive to buy, require lots of work to light and extinguish and clean up after, particularly in large rooms with high ceilings, and produce indoor air pollution, so the first light bulb's nonenergy benefits were dramatic.

Today's incandescent light bulbs are ten times more efficient than Edison's first lamp, and compact fluorescents are thirty to forty times more efficient. Compared to candles, compact fluorescents are over five hundred times more efficient.

My wife and I recently remodeled our kitchen. We designed a lighting system based on commercial lighting equipment that directs most of the light upward toward the ceiling. Most new kitchen lighting uses downlights (known as cans). The main benefit of our approach is that it provides much higher-quality lighting than cans—no shadows, no glare, and the light is diffused so that it is easy to read the labels of boxes stored in the cabinets. With this system, which relies mostly on indirect lighting reflected from the white ceiling, it's easy to read cookbooks even with your back to the light fixture. No need exists for additional lighting under cabinets or above them. (We installed under-cabinet lights, but we never use them.)

The primary reason we like the system is that it provides better quality lighting. Yet it also saves about 95 percent of the energy a conventional approach uses.

4. General-purpose commercial lighting. Efficient lighting is valued mostly for its nonenergy benefits. Lighting manufactur-

ers and designers who promote the highest energy-efficiency technologies market these products primarily on the basis of improved lighting design and quality, which leads to enhanced worker productivity. This is the preferred marketing approach rather than to emphasize the cost savings from energy efficiency, even though these financial benefits are substantial and provide rates of return on the investment in lighting efficiency, which typically exceed 30 percent. So, for example, the newest generation of fluorescent lamps, which allows a 17-percent reduction in energy use compared to the previous generation, is promoted primarily on enhanced-color rendition and extended lifetime.

An office I helped design in 1987 demonstrates how the use of one-half-watt-per-square-foot lighting continues to be an example of green building techniques. Yet what impresses most visitors is how nice it looks and how well it functions rather than how little energy it uses.

5. Heating and cooling. For heating and cooling, increased energy efficiency almost inherently involves improvements in comfort. Thus, for example, energy-efficient windows in homes are marketed as heavily based on their improvements in comfort, noise reduction, and cleanliness (prevention of outside dirt particles from entering the windows) as for energy efficiency.

Superior insulating materials, whether insulation or windows or reductions in air leakage, inherently lead to higher levels of thermal comfort. Consider, for example, a home heated in the middle of winter. The air might be heated to a comfortable temperature, say 72-degrees Fahrenheit, while the air outside of the wall might be at a cold temperature of, say, 10-degrees Fahrenheit. The wall will therefore have a distribution of temperature from the inside to the outside surfaces. This results in a range from slightly below the interior temperature to only slightly above the outside temperature.

This distribution of temperature depends on the level of insulation of each material: the temperature drops more across better insulators than across poorer ones. One component of the effective insulating value of a wall is the resistance to heat transfer between the wall of a room and the air inside. If this so-called thermal resistance is a large fraction of the total resistance of the wall—in other words, if the wall is poorly insulated or not insulated—the wall surfaces will be cold.

Our bodies radiate heat to cold areas, but the cold areas don't radiate much heat back. Therefore people will feel uncomfortably cold near an under-insulated wall or window even though the thermometer may indicate that the air should be comfortable.

In contrast, for a well-insulated wall, virtually the entire temperature drop occurs within the insulation, which places the wall surface temperature close to the room temperature. This creates a feeling of enhanced thermal comfort.

Cooling is similar. During a hot spell, a well-insulated wall will be only slightly above room temperature, whereas a poorly insulated wall will be much hotter.

This is particularly evident of windows where equivalent insulation levels generally are lower than walls (although with the best-performing windows, near-comparable performance is possible).

For cooling, the most energy-efficient windows also will keep out more than half of the sun's heat without any alteration to the window's appearance. Such windows keep out invisible infrared heat from the sun—about half of solar heat is in the infrared, rather than the visible part of the spectrum. A person who stands in the sunshine behind a "low-e" heat-rejecting window will feel much less warmth than the same person who stands next to a conventional clear window that looks identical.[1]

An efficient window's ability to block summer heat from entering the building also has a valuable nonenergy benefit: enhanced fire safety. In the event of a fire, efficient windows hold up longer than ordinary windows, which can allow fire fighters to contain a blaze before it spreads farther or at least can provide more time to evacuate occupants.

6. Air tightness. Even more dramatic advantages are possible

when the air tightness of a house is improved. For example, duct heating-and cooling-distribution systems in homes frequently lose a quarter of the energy that the furnace or the air conditioner puts into them. This can result in some rooms being either overly hot or cold. But more importantly, leaks in the ducts create unexpected pressure differences in the house. In some cases, these pressure changes create suction rather than positive pressure. If one of these areas subject to suction is a furnace room, a room with a water heater or fireplace, or a garage or other storage room containing materials that give off toxic fumes, this can lead to serious problems.

"Negative pressures" in rooms with combustion appliances can lead to the creation of carbon monoxide and its introduction into the air distribution throughout the home. Carbon monoxide is an acute poison and accounts for a number of deaths every year in homes in the United States. When ducts are sealed to prevent leakage, this not only improves energy efficiency, but also reduces the likelihood of indoor air-quality problems and comfort concerns.

Making the rest of the house—the windows, walls, and doors—more airtight also improves indoor air quality as well as saving energy if the house is provided with mechanical ventilation (now required by advanced codes and specifications for energy efficiency). Reliance on random leaks and construction errors to provide fresh air is a poor ventilation strategy: studies have shown little or no correlation between leaky houses and low air-pollution levels indoors. This is not surprising because while a leaky house may provide lots of fresh air when it is windy or cold outside, it may provide almost none under other conditions. In contrast, a fan-based system for ventilation works equally well every hour of the year, and can assure indoor air quality.

7. Location efficiency/smart growth. The nonenergy effects of smart growth are immensely favorable economically, at least when such policies are designed to enhance rather than restrict choice. Research on property values is beginning to sug-

gest that smart growth neighborhoods are more desirable, as reflected in higher property values.

Perhaps one reason for higher property values is greater convenience and time savings, in addition to cost savings. In suburban sprawl, mothers typically spend seventeen full days a year behind the wheel, which includes the time required to drive their children to their schools, friends, and sports events, et cetera.[2] This is more time than they spend dressing, bathing, and feeding a child.

In smart growth areas, children over a certain age can walk, bike, or use transit to reach destinations by themselves. I live in a smart growth area and found that my children could travel unaccompanied either by foot or on transit as of the fourth grade. Not only did this save time and stress for my wife and me, but also the children felt empowered and were happier as well.

Innovation, Process Improvement, and Cost Reduction

Energy efficiency, as discussed thus far, has focused on energy uses within buildings. This is not accidental, because buildings account for more than half of electric peak power and some 35 percent of overall energy use in the United States.

However, similar arguments apply to industrial processes. An engineering project intended to save energy may produce process improvements whose value greatly exceeds the value of the energy savings, even when these savings alone would be substantial enough to justify the project. These process improvements may result in better production or in higher output, lower cost, or greater consistency of production. In some cases, the need for a factory to produce a more efficient product leads to innovations that improve the production process even more than the factory improved the product itself.

Two stories illustrate this concept:

1. In the mid-1990s I visited two plants that produced clothes washers (both at the invitation of the manufacturer).

One plant produced an innovative, new, highly efficient clothes washer, while the other produced conventional, relatively low-efficiency products. The difference I noticed between plants was dramatic: the former plant's premises were cleaner, better organized, better lit, and the workers were evidently happier and had stronger company loyalty. Worker ergonomics on the assembly line appeared to be quite good (the management mentioned this as a specific goal of their new assembly line along with energy efficiency).

At the latter plant, workers' moods clearly reflected some underlying levels of rebellion against the company, and the assembly process clearly involved more discomfort and a noticeable higher potential for repetitive-stress injury to the workers. The workers' attitudes were sloppier and the plant was physically messier and dirtier. The threat of labor unrest was palpable to the visitor. It was also evident that this plant hadn't changed its processes in a while.

2. A staff member at the California Energy Commission found a similar situation when the Commission required a change from old-fashioned metal-framed windows to vinyl-framed windows in order to comply with the energy-efficiency building code. Plants that produced the high-efficiency windows were better organized, cleaner, and more businesslike. Plants that produced less-efficient product had debris scattered around the floor and looked much more marginal.

Doesn't industry, which has market incentives to maximize profit, succeed in investing in new technologies even if residential consumers don't? Unfortunately, the answer is no: even more formidable barriers limit industry's investment in innovative technology that reduces pollution and energy use.

Perhaps the most interesting evidence for this finding—that industry fails to invest in new technologies that will reduce its own costs—is evident as we look at the refrigerator story's next chapter: the remarkable decline in price.

The Steady Decline in Refrigerator Cost

Efficiency wasn't the only aspect of refrigerator performance that manufacturers successfully improved over the last thirty years. As refrigerators became more than four times more efficient, eliminated ozone-depleting chemicals, and upgraded their size and feature offerings, they also simultaneously became 50 percent cheaper.

Figure 9 shows the average cost of a refrigerator as sold by the manufacturer. We see a curve of cost that continually declines, measured in real dollars (that is, in dollars adjusted for inflation), which amounts to a 50 percent reduction in real cost. Remarkably, this reduction occurred steadily in the face of the nearly 20 percent increase in size shown on the graph and regardless of the replacement of manual-defrost refrigerators (which were nearly a quarter of the market in 1972) with fully automatic-defrosting models.

Initially one might be tempted to say that this reduction in cost occurred despite six iterations of energy-efficiency standards and two iterations (mid-1990s and 2001) of ozone-depletion phase-out regulations. Government regulators and even environmental advocates predicted each of these regulatory upgrades would have a significant cost—typically between $30 and $80 per unit. But none of the predicted cost increases was realized.

The results displayed in Figure 9 below are so striking that it seems evident that these reductions in cost occurred because of these regulatory challenges rather than despite them. The need to provide continuous improvement in the environmental performance of refrigerators provided manufacturers with the motivation to improve technologies throughout the production process that supplemented other significant market-based incentives to cut price—namely the emergence of large retailers that bought in bulk and therefore could force manufacturers to be more competitive. The retail competition also led to lower markups to the consumer, so actual consumer prices declined faster than shown in this graph.

We can illustrate the observation that standards and incentives lead to innovation by looking at the details of the engineering analysis that underlay the 1992/3 standards proposal. The California Energy Commission and the Department of Energy had promulgated the standards based on the expectation that specific technologies would be used to achieve the required results in terms of energy efficiency. The economic analysis was based on three variables: how much would these new technologies cost the manufacturer and how much would they add to the retail price of a refrigerator, and what would be the value of the energy savings to the consumer. The standard was based on a set of technologies that all saved more than they cost.

Figure 9
Cost and Energy Use of Refrigerators

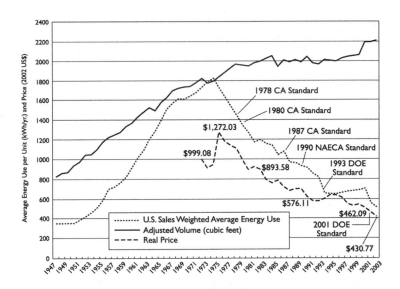

Fortunately the structure of the standards (and subsequent incentives) had been designed more cleverly than that. Manufacturers were not required to use any specific technology; instead, they were required to meet an objective measurement of energy performance for which they could use any means at their disposal. We learned that the 1992/3 standards were met by incremental improvements to existing technologies rather than the more dramatic and sexy improvements that the government analysts and nonprofit organization experts had expected. This is probably one of the reasons why the improvements were achieved without a net-cost increase.

Economic and Environmental Success

What was the effect of these sets of regulations and incentives on refrigerator manufacturers? While it's difficult to offer a precise answer, two things are clear. First, the manufacturers had to work appreciably harder at engineering design than they would have otherwise. But secondly, an examination of the stock prices of those companies primarily engaged in manufacturing appliances over this period indicates that they did pretty well.

Maytag and Whirlpool stock prices generally rose over the thirty-year period in which they responded to environmental regulations and incentives and achieved their remarkable accomplishment in improving energy performance and consumer attractiveness for ever-lower prices. It depends on how you select the starting and ending points (stock prices are volatile) but as an approximation Whirlpool is worth some six times what it was prior to the energy regulations circa 1975, while Maytag is worth three times as much. Certainly these appliance manufacturers did substantially better over the last fifteen to thirty years than auto manufacturers, who were exempt from increased energy-efficiency requirements after the mid-1980s. Ford and General Motors, the two major car manufacturers whose stock prices are accessible, saw no appreciation at all in their share prices from the mid-1980s to 2005, and sharp declines thereafter.

Of course, Whirlpool and Maytag make many other products besides refrigerators, so the actual situation is much more nuanced than this.

While we thus far lack sufficient information to draw definitive conclusions, the argument that the energy-efficiency standards benefited everyone is certainly plausible and worthy of careful study.

Cost Reductions for Energy Efficiency in Buildings

American appliances is not the only area where mandated improvements in energy efficiency led to cost reductions instead of the expected increases. In Russia most buildings were constructed with little or even no insulation—even into the 1990s—despite the cold winters. I worked with Russian officials over the 1990s to develop and implement new energy codes for buildings that were adopted regionally beginning in 1994 and became a national standard in 2003, which produced a savings of about 40 percent in heating energy.

Russian buildings were constructed of large factory-made wall panels that are assembled and attached together at the building site. (This construction method was designed to allow a year-long construction season because most of the labor could be undertaken in heated factories rather than at the construction site.) The panels usually consisted of a sandwich of insulation between concrete, but the insulation material was pierced by numerous cross connections that allowed heat to short-circuit the insulation.

To meet the new standards, wall panel manufacturers increased amounts of insulation, or added it when none had been used, and paid more careful attention to the heat flows through structural members next to the insulation. These actions raised the overall insulating value of the panels by about threefold, while the dimensions of the wall remained the same in order to reuse the same concrete forms at the panel factory. This was expected to add modestly to the cost of the panels.

However, it turned out that when the change had been made and modern insulation materials came into wider use, the cost of a cubic centimeter of insulation material was less than the cost of a cubic centimeter of concrete. Because the forms stayed the same size, the net effect of the standard was to replace concrete with polystyrene insulation. This switchover resulted in lower costs.

So the net cost of the new standards was less than zero.

Overcoming Barriers to Innovation

A rather interesting illustration of process improvements whose value would have greatly exceeded the value of the energy savings had it not been prevented by a company's failure to innovate is a project the Natural Resources Defense Council, a national environmental advocacy organization, performed in partnership with Dow Chemical.[3]

Dow Chemical has a reputation as one of the best-managed industrial companies in terms of its use of up-to-date technologies to improve efficiency. But in one particular city, local environmental activists protested Dow's emissions of toxic chemicals from the local plant. The Natural Resources Defense Council got involved at the invitation of the activists and suggested so-called waste minimization as a means to address the problems. Waste minimization is a method by which an industry changes its industrial processes so that it emits less while providing the same or improved product output.

In the case of Dow, the Natural Resources Defense Council suggested that Dow make use of an outside consultant who could point out the opportunities for profitable waste minimization, provided that Dow agreed to implement the results of the consultant's study. The company initially resisted this suggestion and claimed that no cost-effective options existed for reducing emissions.

After several years the environmentalist coalition and Dow agreed to accept the results of the consultant's study. The consultant actually found numerous opportunities to cut waste. The final project led to a 43 percent reduction in toxics and a return on investment of 180 percent for the company.

This would seem like an obvious win-win situation. However,

Dow failed to implement similar processes in any of its other plants, and no other chemical company has worked with an environmental organization (or by itself) to achieve a similar success story. Indeed, the company's reaction was that the level of increased profitability generated by the project, while high in relation to the capital investment involved, wasn't worth the trouble to replicate.

But how much trouble could that be? If a project saves over $1 million a year, how could one argue that to hire an engineering manager to oversee similar projects throughout the corporation in the future is not worthwhile? The Natural Resources Defense Council scientist, Linda Greer, subsequently had a similar experience with another company where, again, a very economically attractive waste-minimization process was discovered and implemented. Yet after the completion of the project, neither the participating company nor its peers made further progress.

Dr. Greer realized that to replicate these two successes would require an army of environmental advocates to knock on the doors of every industrial plant in the country and to continue to work with its managers until they finally agreed to implement such projects.

Utilities that operate industrial energy-efficiency programs frequently find the same results. A project intended to save electricity in an ice-cream factory in New England also discovered ways to increase the productivity of its staff. The cost reduction was so great that the parent company, which previously had planned to shut the plant and lay off its staff because of uncompetitive high costs, was instead able to retain the plant in operation. This project helped the region's economy far more than would be expected for the size of the energy savings alone. This is not an isolated example, but rather a repeat story throughout the country when utilities look in detail at the opportunities for enhanced efficiency in their customers' industrial operations.

One common excuse for choosing not to make these investments is that the company has a high "hurdle rate" for investments: that proposed investments must jump over a hurdle of 30 or 50 percent annual returns in order to be able to compete with other attractive company investment options.

But such an argument is nonsense. If a company has a hurdle rate of 30 percent return on investment, that means it only invests in options that will return more than 30 percent. But if the corporation uses this hurdle rate for ten or twenty years, it means all of its capital is invested in greater-than-30-percent annual returns, and the company itself must be returning over 30 percent to its investors.

However, no company in the world is consistently doing this well. Actual returns to American shareholders are typically 6 percent or less, after inflation.

How Business Ignores Process Engineering

As part of a study on industrial energy efficiency in the United States, Professor Marc Ross of the University of Michigan looked at, among other industries, the American steel industry. He found that typically a production manager who ran a $100 million/yr assembly line had no access to engineering staff whatsoever. The point of this observation was that if potential innovations to save energy were technically available, they would always be unknown because the manager lacked the staff expertise to even look for them, much less find or implement them. This is a fairly typical pattern.

Interestingly, when Japan decided to address the nation's energy problem after the energy crisis of the 1970s, it required all industries over a particular size to hire and maintain on staff an energy manager whose job was to look for energy-efficiency opportunities. By all measures Japan's industrial energy efficiency improved and was at world-class levels within a decade.

How would an industry address the issue of how to find and implement investments in energy efficiency or emission reduction? The broadest and most profitable approach would be something called process engineering. Here one must do more than merely improve the efficiency or environmental characteristics of individual components of a fixed process, one must also look at changes in the entire process to minimize waste and maximize productivity. One would think that in such a technology-driven economy as the United States process engineering would be a major field in the industrial sector, yet actually it's the reverse that is true.

One international consulting firm that has worked with major U.S. companies on energy-efficiency issues for the last several years has found a similar set of reactions. Most companies look at energy efficiency or other environmental improvements as something that must be done only once and under outside pressure. No internal program, comparable to quality assurance or labor productivity-improvement programs, exists to provide continuous feedback regarding how energy-management processes work or how to locate new opportunities.

The consulting firm mentioned above finds and advises its clients that the issue is primarily not a technical problem, but rather a managerial problem: how to imbue within the corporate structure a sense of the need for continual improvement with respect to energy use and the assignment of individuals responsible to direct such a program on a regular basis. Particularly few of its clients in the United States have implemented this program to the extent that would make sense. Apparently, however, the receptivity of foreign companies to accept this sort of advice is somewhat greater. This poses an interesting challenge to future U.S. economic competitiveness.

Here's a personal story that illustrates the issues mentioned above. My wife entered the work force as a chemical engineer in the late 1970s and was immediately hired by a local plant that produced pigment for use in concrete. Fresh out of college, her job was as a process engineer. Under the direction of a senior engineer, her first project was to change the fundamental process that converts the raw material (primarily iron) into the product, which is a hydrated iron oxide.

The fundamental reaction is rather similar to how iron rusts: it releases rather than requires energy. Therefore, evidently large potential benefits in energy efficiency emerge if the plant were to change the previous process which involves the addition of heat in order to make the reaction work.

My wife and her colleague developed a new process that used mechanical agitation of the iron rather than relying so much on adding heat; it reduced energy consumption six-fold while it also increased throughput (which was even more important economically to the company than the energy savings) and product consistency. After they looked at the efficiency of the entire process, her department was able to provide far more energy savings than a specific focus on energy would have found, and it was also able to produce much larger benefits than just the reduction in utility costs.

The main goal of process engineering is to look creatively at what the nature of a problem is and how it can be solved comprehensively. One particular company's attempt to solve a problem with an industrial dryer is a good example of this successful approach. In this facility, the immediate problem that attracted managers' attention was that the dryer performance was inconsistent and therefore unnecessarily expensive. The managers hired a process-engineering consultant to fix the dryer. The consultant initially considered a very simple approach to how to solve the problem—improve dryer performance—but after a more detailed examination the consultant realized it wasn't actually necessary to dry the product at all: the less-than-dry product could suit customer need just as well as the dried one. So rather than solve an obvious and perceived small problem, the engineer ultimately solved a more significant one, much more comprehensively. This solution was much more profitable than it would have been to solve the immediate problem of the poorly performing dryer.

The real policy issue is much deeper than a couple of stories about remarkably successful projects in process engineering; it is a problem of industrial organization. The key question is: why did my wife's plant have a senior engineer with the authority to work on process engineering full time (with a two-person staff)? To answer this question, we must note that the senior engineer was a highly valued, long-term employee who previously had been plant manager. Management was not his forte, so the company replaced him with someone who was more management oriented, but wanted to retain his expertise for use in the plant. The company created a senior project manager position and turned him loose on projects that, at it turned out, more than repaid his full salary and that of his staff.

The above is an unusual circumstance in America, however. More typically, my wife found, process engineers were hired fresh out of college and served a type of apprenticeship in the process-engineering department where, if they were any good, they were promoted to production managers. So who are the process engineers? If they exist at all, they are young college graduates or people who appear to lack the talent to perform "real" production management. Clearly they are not the ones who will be most trusted or consulted in decisions affecting plant operations.

If anything, this situation has worsened in the past two decades. A search for "process engineering" on the Internet reveals references to minor chapters of obscure books, references to a smattering of university and trade-school courses, and a few firms (mostly smaller firms) that offer process-engineering services (but usually as one among dozens of options). In other words, process engineering has yet to become a serious part of the activity of managing the American industrial enterprise despite ample evidence (consider the Dow example) that rather attractive efficiency options for industrial managers exist if they only look. And even the process engineering that does exist may well be regulatory-driven rather than designed to maximize profit.

One reason corporate CEOs cannot implement a wide range of technology innovations, and particularly those that would improve the environment, in their operations is because they have no one to advise them on what the opportunities are. Indeed, so little process-engineering expertise seems to exist in many companies that even to outsource the work to consultants is impossible because of the lack of staff available to write a professional quality scope of work.

This is a serious problem, not only for mid-term corporate profitability but also even more for long-term economic competitiveness. For the past century, the competitive advantage of American industry has been technology. American industry was generally more productive than that of competing countries.

However, in the twenty-first-century business environment of rapid technological change and the diffusion of information globally, productivity advantages can be short lived. A company, or an industry, that is unaware of the potential for continuing engineering improvements that get more profit out of a given level of energy use, or staff time, or materials consumption, risks the loss of its competitive edge.

Yet most companies seem willing to court this risk—remarkably few companies act on the principle of continuous improvement. Few American companies even monitor their energy use to measure possible fluctuations. A company cannot manage efficiency improvements if it lacks even the ability to measure them.

How Environmental Regulation Promotes Process Engineering

Such lack of managerial emphasis on efficiency explains why we need policy interventions to realize the potential for new technologies to contribute to economic growth and environmental improvement. Evidently, to obtain long-term economic growth, we either need new management philosophies or a continuation of policy involvement because even if a given policy successfully adopts the best practices for 2006, the same institutional failures will impede the introduction of next-generation technology.

Why are environmental policies needed to promote the type of innovations that are developed through process engineering?

The answer is that usually the innovations appear to affect only a small part of the business cost structure, so they seemingly warrant low priority. Energy costs typically account for only 3 percent of total operating costs in industry, so there is no reason that the energy/environmental issue should get top management attention no matter how high the rate of return on the investment.

What causes management to pay attention to the energy/environmental issue is a regulation or a strong incentive. If the company cannot continue to produce its product without changes, that's a high-priority issue that will attract sufficient resources to address it. If it's merely a matter of a little extra money, it could be put on the back burner.

Strong incentives provoke the same reaction. If the failure of the company to develop a product that qualifies for an incentive results in a major loss of market share to a competitor that offers such a product, this too will get management's attention.

The problem is that when management ignores high rates of return in one area, such as energy, it will ignore other high-profit opportunities in alternative areas for the same reason. So while any one opportunity may affect only 3 percent of costs, the entire portfolio of ignored opportunities is substantial. And the failure to look for them can be a problem itself—one that leads to a broad pattern of technological stagnation.

Numerous stories exemplify how industries can slip into patterns of stagnation when markets become relatively uncompetitive, yet can be spurred to innovation when more competition is induced, either by regulations or policy or even by other causes. For example:

1. The Telephone—Consider the telephone as a simple everyday example. A new mobile phone purchased in 2007 will have an electronic phonebook that allows five hundred or more entries, a PDA with alarms for scheduled events, a digital camera, downloadable ring tones and games, and recall memory of the last ten or more calls received and sent; many offer tape recorders, Global Positioning Systems to pinpoint the location of the phone and allow web-based downloads of information on nearby restaurants and movies, and the ability to download,

store, and play music (and other features), all within a box that weighs less than four ounces. In contrast, desk phones still work more or less as they did thirty years ago, save for the cordless feature. Why the difference? Simple: wireless phone service is much more competitive.

When industries slip into less-competitive operation, technological stagnation results. This applies particularly for difficult-to-see features like energy efficiency.

Only a limited number of firms dominates many of the industries whose operations most affect the environment. So if the corporate culture of only four or five of these companies ceased to support innovation—and a consultant on energy efficiency notes that only a select few companies actually do support continuous improvement of technologies in their own operations—no significant impetus for anyone to change would exist.

A group of four or five industry leaders could easily develop management methods such that when they slipped into stagnant patterns none would notice it; and without other large and more innovative competitors, nothing would upset this arrangement.

Americans are an optimistic people. We tend to believe naturally that technological progress occurs automatically and that it makes all things better. Thus a built-in assumption exists that energy efficiency naturally increases in the free marketplace.

However, such an assumption is not necessarily the case. Instead, we find many examples where a lack of policy intervention leads to stagnant technologies, a situation that does not change of its own but is only corrected by the initiation of environmental policies.

2. Water Heaters—Water heaters are one of the largest uses of energy in the home. Yet between World War II and the mid-1970s, the efficiency of water heaters actually declined somewhat. Standards adopted in the 1980s brought the nation back up to World War II levels of efficiency in water heaters, and regulations promulgated in and around 2000 finally have led to

(modest) improvements in technology, both in fire-risk reduction when the water heater is exposed to flammable fumes and in terms of improved energy efficiency. But we still aren't back to World War II levels of environmental efficiency because an increasing fraction of water heaters employ electricity, with an emissions level more than double that of gas-fired water heaters (and consumer costs almost double with the switch to electricity). Newer technologies are available that can cut electricity use in water heaters by about two-thirds, but they are not widely used—another illustration of how difficult it is to promote cost-effective efficiency technologies.

3. Transformers—Virtually all the electricity used in the world passes through at least one transformer as it travels from the power plant to the end use. Utilities own most transformers. The basic technologies of energy efficiency in transformers are well known. Over the past decades, new methods to manufacture the steel used in the core of the transformer have increased the efficiency potential of transformers. Yet the average efficiency of the transformers utilities use has declined since the early 1990s, as utility regulatory changes have made it impossible for many utilities to recover the additional costs of the more efficient transformers.

This result is extremely ironic because the regulatory changes were intended to enhance market forces and competition. Instead, utilities that used to invest in transformer efficiency following the normal business ideal of minimizing costs over the life of the product began to buy transformers that minimized initial investment. The forgone efficiency measures would have had more than a 35-percent return on investment, compared to typical utility returns below 10 percent.

4. Windows—Window energy-efficiency technologies were similarly stagnant until the mid-1970s when energy codes required higher levels of performance. Up until the 1970s, energy efficiency in windows required two separate windows, or storm windows; or the use of sealed, double-pane glass units. None had energy efficiency beyond what was available in the

seventeenth century, when European windows accommodated a two-pane glass design.

In the 1970s, with U.S. government-funded research support, manufacturers developed low-emissivity coatings (see "low-e" described in Chapter 3) that reduced the transfer of heat by radiation from one pane of glass to the other, suspended transparent films that offer additional opportunities for low-e coatings and the presence of a nearly invisible third pane of materials, and inert gas fillings that reduced heat conduction. Other research established how the window frame affected heat loss and gain, and showed how metal-framed windows often fell below expectations because the heat flow through highly conductive metal short-circuited the insulation of the glass portion of the window.

Yet none of these research developments made much difference in the market until state energy codes demanded improved thermal performance in windows. Such codes were adopted in the early 1990s.

As a result, it is possible to buy windows today with a resistance to heat flow of between R-6 and over R-10. (Resistance to heat flow is measured in degree days Fahrenheit-square-feet per Btu per hour and is called R-value. Higher is better.) The performance of this new technology can be compared to an R value of about R-1 for a single-pane window, R-2 for a good, old-technology double-pane window.

Windows at R-3 and better are nearly universally used for mass-produced products for new construction in colder U.S. climates. Their incremental costs compared to double-pane windows are essentially zero.

The success of these products was due to highly directed government programs to encourage efficiency, not to normal market forces. These programs began with government-funded research and development followed by economic incentives, the development of energy-efficiency codes, and such normative rating systems as Energy Star® that direct consumer attention toward highly energy-efficient products.

As a result of the innovations that came as low-e windows with thermally improved frames were produced in bulk for several years, the incremental cost of going from an R-2 window to an R-3 window, which was estimated at several dollars per square foot in the 1970s, declined to the point where at least one major window manufacturer doesn't even offer the lower-performing, non-low-e-coated double-pane glass because the difficulty to stock two different lines of windows overshadows the savings (in manufacturing cost). Recently the same set of government interventions has introduced into the mass market the solar-reflective window, which uses a different type of low-e coating to reject unwanted solar heat in the summer, as well as provide better thermal insulation.

5. Refrigerators—A good example of an innovation-inducing set of regulations is the efficiency standards for refrigerators, particularly those California adopted in 1984 for effectiveness in 1992. Recall how effectively these standards achieve their direct goal to reduce energy use. Let's now examine how such standards, in order to achieve the goal, encouraged innovation with unforeseen designs that cost less and performed better than expected.

The California standards were performance-based in that no particular technology was required. Instead, the standards set a required maximum level of energy consumption per year, measured in an easy-to-perform laboratory test whose results could be simulated by computer or actually tested on a physical refrigerator prototype. (The standard depended on size and features, so that a bigger, more feature-laden refrigerator would have an easier—higher—energy target to meet than a smaller, simpler one.)

The standards were adopted in 1984. In 1986 I was seated across the negotiating table from refrigerator manufacturers where we discussed the prospect to extend the California standard nationwide through mutual agreement. The manufacturers resisted this extension.

The manufacturers persisted in their initial claims that the level was basically impossible to meet at a reasonable cost. I disagreed and explained why I thought they were resistant. The people with whom I negotiated were the government-affairs representatives of the refrigerator companies, not their engineers.

I told the manufacturers that they probably had asked their engineers what it would take to reach compliance with the standard, and that the engineers had come back with something like: "I'm not sure exactly how we could do it. We had this initial idea and we tried to build a prototype and it didn't work as well as we expected so we still don't know what it takes to comply, or whether we can even do it competitively."

My negotiating partners' eyes practically popped out of their heads. They apparently thought that I had bugged their internal communications. What actually happened was that I was able to describe the normal process of development that would occur when manufacturers seek technical innovation.

I explained further that few designers ever get it right the first time. An engineer who is responsible for the investment of hundreds of millions of dollars in product improvements will only claim the problem is solved when he or she is 100-percent certain that it is, in fact, solved. So the engineer will undertake a particular approach that looks most promising and, usually, find some flaws that exist with parts of the initial conception. This person will then return to the drawing board and try a second-generation approach, which usually works better, but may not get all the way to the solution. With seven years to fully implement the solution to the 1992 standards, and at least five years to design the prototype (the last two years are to develop the actual manufacturing process), any firm answer developed with confidence after only a year and a half to two years would almost certainly be problematic. Ultimately, with manufacturer A fixed on that approach, it would seem apparent that manufacturer B, who would continue trying to make further improvements, would be competitively stronger.

However, the reaction I received in the negotiation illustrated pretty clearly that while the California Energy Commission was debating over the level of the refrigerator standard, the manufacturers had no idea about how they would actually comply and no process in place to evaluate how to make further improvements in energy efficiency. Even eighteen months later, they still didn't know what they would be able to do.

Indeed, it would have been foolish for them to know. If the market failed to respond to available opportunities to save 20 percent or more of energy use with a 25-percent-or-more rate of return on investment, why spend real engineering resources on doing even better? The results of such an inquiry, even if it were successful, would fail to produce increased sales or increased profits. So the process, as it looked in 1986, was that manufacturers were playing a game of "catch up" on efficiency engineering. They were unaccustomed to serious engineering challenges of this nature, and so it unsurprisingly made them nervous.

Nervousness may or may not have been an appropriate management strategy for the engineers or the production managers at the time, but the results of compliance to the standard clearly showed that the innovations induced were impressive indeed. When the 1992 California standard was enforced nationally (in 1993), manufacturers not only supplied models that complied with this ambitious standard at no additional cost, but also actually had significant numbers of models that were 10 and 15 percent beyond the standard (in response to utility-based incentive programs).

Manufacturers beat the ambitious energy goal and accomplished this with a better choice of technologies than any of us had predicted. While the standards were performance-based and required no particular technology, the level of performance demanded was based on the assumption of specific technologies. This is how the regulators could predict that the energy savings would pay for the increase in product cost—a requirement of the appliance standards law.

When the 1992 products were introduced, they didn't use the technologies that the regulators or the standards advocates had expected. Instead, the manufacturers found a large number of incremental improvements they could make to existing products—technologies that cost much less than we had expected. Indeed, when we look back at the cost of compliance, the data indicate no cost at all occurred. As shown in Figure 9, refrigerators that met the advanced energy standard actually cost a little less than they did before.

The difference in energy efficiency was so dramatic that some retailers marketed the concept that customers should replace fully functional old refrigerators with new ones in order to reduce their utility bills. This marketing pitch made sense because a consumer in an area with high electric rates might pay $300 annually to run an older side-by-side refrigerator. A consumer who buys a new one could actually pay back the entire new purchase price based on the electricity savings alone within just five years.

6. Lead Phase-out in Electronics—Opportunities to induce new investments whose returns fully compensate for the costs of compliance apply beyond just the energy area. An article in the *San Francisco Chronicle* that addressed the efforts of industry to comply with California and European Union regulations that banned the use of lead in electronic components notes, while compliance is estimated by industry to cost $5 to $10 billion annually, this need not be a burden.[4] One manufacturer stated, "While you're doing this [investing in compliance], you might as well re-design for better cost efficiency, phase out old products and rationalize the number of suppliers." So the regulations are actually "creating a whole new industry."

7. State Clean-Air Regulations—Similar stories exist with respect to clean-air regulations. When the California Air Resources Board needed to crack down on certain categories of industrial emissions in the Los Angeles area, both of the major investor-owned utilities worked with their customers to allow them to comply with the new clean-air regulations while they retained their jobs in the Southern California utilities' service territories. It turned out that to paint and to coat products pro-

duced significant amounts of smog and caused worker health-and-safety issues in the spray booths. New technologies were developed that use techniques like electrostatic attraction to apply coatings directly from the source to the work and avoid having the spray reach all over the room. This was a far more efficient use of the coating product and also produced better, more uniformly coated products—and reduced smog and workers' exposure to solvents dramatically.

Improvements such as those mentioned above allowed the Los Angeles region to retain jobs that likely would have been lost to other regions simply due to economic competition pressures even without the air-quality regulations. In other words, the air-quality regulations helped improve productivity so much that the jobs were retained.

National Economic Development Policy and the Environment

Virtually everyone supports economic development. If a policy could be articulated that produced more jobs and made people wealthier, everything else being equal (including the environment), it is difficult to imagine that anyone would oppose it.

Yet despite the appeal of economic development, we know precious little about how to encourage it, whether on a national, a regional, or even on a neighborhood level.

Globally, immense differences in levels of economic development and their rate of change have existed over the years. For example, Japan, Korea, and Hong Kong vaulted from poverty to high levels of economic development within a few generations, while in a few cases, such as that of Argentina, countries fell from high levels of economic levels to much more modest circumstances in less than a century. China's economy currently grows at 7-percent annual rates or higher, while the economies of most African countries remain stagnant.

What explains these differences, and what can we do to foster economic development, either in the United States or in other countries? The few general principles we have observed as we compare different countries are not specific enough to be of much policy use. We know that economic development is fostered when a country attains:

1. High levels of education
2. The rule of law provided by stable governments
3. Higher levels of private investment

Yet these observations are difficult to translate into policies that can be implemented. Even within the United States, vast differences exist in economic development from city to city and between neighborhoods within the same city. Decades of effort to advance the progress of depressed areas have failed to yield breakthrough successes.

This section explores some additional reasons why environmental policy could be the most important driver of economic development and growth in practice. It may not be the most influential factor overall, but it is certainly one of the most controllable ones: we know quite well how to promote energy efficiency through state and national policy initiatives and how to improve air and water quality to assure land preservation, and how to secure other key environmental benefits. In contrast, prescriptions on how to improve education or enhance rule of law are much more challenging to find and even more difficult to implement.

Access to Resources

The broad issues of competitiveness and national economic-growth policies have for decades been discussed in the context of regional planning. One of the most interesting observations was about the importance of resource efficiency—a concept that closely parallels strong environmental policy—in that it allows economic development, even for countries that start out poor and technologically backward.[5] Because cities and their regions in developing countries do not have access to local production of the basic raw materials and resources needed to grow an economy, resource efficiency is an imperative. All locally used resources have to be imported at great expense, so a city/

region that minimizes its use of these resources can maximize the amount of its expenditures that stay in the economy and contribute to development.

If significant use of natural resources is not available as a source of economic development, what can substitute? The obvious answer is innovation. Innovation-friendly policies have been a part of the development strategies of countries like Japan, South Korea, Singapore, et cetera: education, diversity (the ability of people and organizations to learn new ideas from those of other regions, ethnicities, or cultures), and organization (which also fosters the ability of individuals to learn from each other and innovate). Better information transfer and people's ability to share it substitute for brute-force use of natural resources.

Pursuit of such policies that focused on human capital and resource efficiency rather than brute force led to economic performance that outran that of better-endowed countries. And countries that have demanded an increase in environmental performance generated more competitive business organizations.[6]

Conversely, we now see evidence that natural resource exploitation hurts economic development.[7] A systematic comparison between different countries illustrates "the lower the share of exports of natural resource [as a fraction of national economic output] the faster the rate of economic growth..." Even worse, the higher the reliance on natural resources in the economy, the greater the rate of income inequality that is observed. Thus given the choice of policies to develop natural resources and use them freely and policies to refrain from resource development and use present resources more efficiently, efficiency is better for growth. It is also better for the poor.

Human Capital

Environmental protection may also promote growth in an indirect yet important way: to attract creative people who can develop innovative businesses in a particular region or country.

Human capital, and particularly the economic value of creativity to a region, is a newly appreciated factor for growth.[8] Creative people whose work is unrestricted by a particular factory or other location have much greater freedom to live where they choose. They will choose places where it's comfortable to live. While the importance of environmental quality to where creative people choose to live has not been studied in detail, the current research illustrates that an individual with a choice to live in a place with clean air and easily accessible recreational open space or a place with the reverse set of circumstances— that is, a creative person who has a choice of job location and who will contribute to economic growth of the region that she chooses—will choose the former.[9]

Creativity is an important issue for another reason. The strong version of the economic growth proposition implies that we can substitute creativity for raw materials and accelerate growth. Creative positions pay substantially more than average work.[11] So not only will environmental policies produce a greater number of jobs, which accelerates productivity growth and diverts money from low-job-intensity occupations (like energy supply and resource extraction) to more job-rich businesses, but also a higher quality of jobs.

The strong form of the proposition that environmental protection enhances economic growth stands on several important observations. First, we find that many environmental technologies and most energy-efficiency technologies that would have failed to develop without the push of environmental needs have additional benefits whose value exceeds that of the direct benefits.

Second, we recognize that environmental policy promotes innovation that improves productivity and opens up new opportunities for progress that no one could foresee. The benefits of the process of developing innovative solutions may be as important as the specific solutions because the process can be applied to other economic needs.

Third, we find that resource efficiency appears to work as an economic development strategy that is easier to adopt than the other factors that seem to promote regional economic development.

The idea to use environmental policy as a method to promote innovation and thereby increase economic growth and job creation would seem like a policy that everyone could support. As discussed, the facts that describe some of the environmen tal initiatives that have had the largest effect demonstrate the positive economic value of environmental protection. Yet most environmental initiatives in the United States are controversial, and much of the discussion about the economy and the environment consists of arguments about how environmental rollbacks are needed to boost growth.

The next section discusses how theory gets in the way of a political debate that is based on these facts. Theoretical considerations drive the policy debates, so it follows that Part 2 focuses on economic theory and the environment.

Part 2

Environmental Protection, Economic Barriers, and Economic Development

Economics is at the heart of most environmental problems. All of the critical environmental issues involve such debates as how much money to spend on pollution control, habitat protection, or other environmental goals, or how to control people's impact on nature.

Economics is a policy-related science—one whose objective is not only to describe and predict phenomena in the real world, but also to use that knowledge to improve public decision-making. So a discussion of economic theory and how it is used and abused in the policy and political world is at the heart of the question of how environmental protection policy can promote economic growth.

Part 2 explores the issue of how to increase economic growth with environmental protection in two unique ways, each of which focuses on the limitations of conventional economic theory, and in particular how it is used in environmental debates.

The first perspective is the misapplication of economic theory for political purposes—the use of economic fundamentalism as a political force. In Chapter 4 I define economic fundamentalism, explain why it is wrong, and illustrate how it misdirects the political and policy discussions.

Chapter 5 presents a case study of one of the worst examples of such misdirected political decision-making: the California electricity crisis of 2000. The study illustrates how economic fundamentalism caused the conditions that set up this crisis, and provides some specific examples of real-world problems whose potential to occur was ignored in the California experiment to create so-called free markets for electricity.

Chapter 6 introduces a second perspective, and looks at how a more detailed and accurate development of economic theory can explain why the private sector is making environmental choices in an inadequate way—how the failure of markets to behave as naïve theory would expect will continue to restrain growth. If we can understand how real markets differ from ideal markets, we can solve the problems and increase growth.

I offer a number of theoretical reasons that explains the fact that so many unexploited opportunities exist to accelerate growth by protecting the environment. The second perspective finds several pervasive failures of markets—failures that systematically discourage innovation and new technology. Part 3 offers a promising discussion of how these problems could be solved.

4 | Economic Fundamentalism— The Use of Economics as a Religion Rather Than a Science

Neoclassical microeconomics provides an intellectually satisfying and complete theoretical basis to understand how markets work. With a few simple and straightforward assumptions, an economist can reduce complex problems to relatively simple mathematical equations or graphs and can determine theoretically the right answers.

However, for any science to be valid, the assumptions must be scrutinized and tested to assure that they are appropriate for the specific problem. And predictions of the theory need to be tested against real-world data. Economic fundamentalists often forget these steps, and talk only about the conclusions, particularly the conclusion that ideal free markets produce the best economic outcome.

In this chapter I offer a detailed look at how economic fundamentalism has thwarted environmental protection and misrepresented environmental policy initiatives based on a series of mistaken assumptions. I describe what some of the most critical assumptions are and show how frequently they're incorrect.

What Is Economic Fundamentalism?

I use the term economic fundamentalists to describe advocates who focus on one narrow and highly conditional conclusion of economic theory as if it were a universal revealed truth. Religious fundamentalists typically look for revealed truth in sacred documents or in oral histories of their religion. But economic theory is very different from any religion; indeed, most major religions regard the basic goal of economics—to satisfy material desires—as inferior in spiritual value.

No major religion (or even minor religion) addresses marketplace issues, other than to require that businesses deal fairly, honestly, and justly—which, ironically, is quite closely related to the concept of government regulation.

Economic fundamentalism is based on a deep misunderstanding of basic economic principles. Economics, as a science, asserts: "If A, B, C, D... are all true, then X is true." Economic fundamentalists twist this proposition into the statement: "In all cases, X is true." In this case, "X" means the proposition that the American economy functions perfectly today and any government intervention will automatically make things worse.

Economics as Science

Economics is a science, not a religion; science is based on logic that takes us from premises to conclusions, and upon observation to test whether the premises and conclusions are correct. Economic fundamentalism is bad economics as well as bad policy advice because its advocates fail both of these tests of science.

One of the reasons economic fundamentalists succeed in their advocacy is that the assumptions of economic science are seriously underplayed or even forgotten in most common expositions of economics.[1] Most economists seem to forget about the assumptions their theories require: they seldom spell them out explicitly in a single place.

And in some cases, economists don't make their assumptions explicit; rather they take them so thoroughly for granted that they don't even realize that they are making assumptions at all.

This is a problem because if even the teachers of economics understate how much their conclusions depend on the assumptions, students—particularly business and political leaders, who once studied economics in college, who are the subjects of economic fundamentalist advocacy—are likely to overlook them as well.

Assumptions are necessary in any science because in the real world behaviors are complex. If we wish to model how people behave in economies, we have to make some simplifications in order to make the job tractable. Economists expect that while the traditional models are oversimplified, they still give the correct answer, or approximately the correct answer, and this often happens. However, in the area of environmental analysis, oversimplification usually fails.

For example, oversimplified economic models predict that no unexploited opportunities for energy efficiency exist. Based on the results given in Part 1, this prediction is wildly incorrect.

We would expect that after at least thirty years of research on energy efficiency conclusively showing that opportunities that shouldn't exist according to theory do, in fact, exist, that economists would look at the problem in detail and adjust the theory accordingly. Such comparison of fact and theory is what distinguishes science from mere opinion.

Compare Facts to Theory

It's essential to compare facts to theory for science because all scientific theories are oversimplified to some degree. If the theory is usable, the extent of error from the oversimplification will be insignificant. But without comparison to fact, the weakness in a formulation of theory will remain unknown. Theorists can go along their merry way for decades and produce papers and recommendations that are dead wrong but they don't know why. Worse yet, other theorists will study the work of preceding theorists and believe the work is accurate, only to then derive further theories based on the earlier ones.

The later theories above will also be wrong, but in a less obvious way. Such theories will follow logically from the earlier theories, and the reviewer will have to remember to discount them as well when the errors of the earlier work are eventually discovered.

The examination of economic theory in this chapter is informed by data on the performance of real markets, primarily in energy efficiency and other environmental attributes. I discuss what assumptions are necessary for the principle that markets necessarily produce optimum results for everyone to be right, and find that not just one or two, but virtually all of them are violated when it comes to the analysis of environmental problems.

As a result of the systematic failure of mainstream economists to evaluate the data on energy and environmental markets, policymakers misapply and take as received truth theories that are patently and obviously contradictory to fact... to the disadvantage of both environmental protection and prosperity. And seemingly no one looks at the theory and questions why it fails.

A small number of economists has for years pointed out how the simple form of economic theory has been stretched beyond its limits when applied to environmental policy. They have identified in detail the ways in which even within the basic economics model of the world, many environmental policy recommendations are based on oversimplified and incorrect application of economic theory and analysis.

Yet these critiques are not generally a part of the policy debate; indeed, they are often overlooked even in economic analyses. I draw on several of these critiques throughout this book. Like these critiques, I accept the basic economics perspective on the world, but I look at what assumptions are necessary to make the models work, and what happens when they are violated.

Why is economic theory so crucial to a discussion of environmental policy and economic growth? The answer is because almost all of the political opposition to environmental protection is justified by its proponents based on economic theory. To show how and why this theory is wrong supports the main conclusion of this book: that environmental policy can promote economic development.

How Economic Theory Serves as a Political Force

Economic fundamentalists put out a simple message: Free markets without government interference produce the best possible results for everyone. Economic fundamentalists argue that solid science and theory illustrate that markets and competition, rather than government control, provide this happy result.

So economic fundamentalism becomes a political reason to oppose environmental policy, particularly when the policy involves government.

The Fear of a Soviet Style Economy

Economic fundamentalists have been particularly successful as they put forth a simple story that Communist economies, which have been much less successful economically, rely on government involvement and government mandates as the driving feature, while competitive markets have a near-total absence of government involvement.

The economic fundamentalist's discussion about the level of government involvement in the economy particularly refers to the issue of regulation: government regulation is portrayed as a Soviet-style type of policy, and the lack of regulation is presented as the ideal for a market economy.

Yet some significant level of government involvement is necessary for markets to exist at all. Additionally, continuous government action is needed to maintain the conditions required for strong competition to exist in free markets.

And even with continuing government interventions—even when markets are as free and competitive as possible anywhere in the world, the key assumptions of economic theory break down when applied to environmental issues. The simple belief that free markets always provide the best answer is wrong.

Here is a common economics joke about a young boy and his grandfather as they walk down the street. The boy sees a $20 bill on the sidewalk and shouts excitedly at his grandpa that he wants to pick up the $20 bill. The grandfather chastises him, and says: "Don't bother to reach for that $20 bill—it isn't real. If it were, someone would have picked it up already."

The story illustrates one of the simplest and, for the purposes of environmental analysis, least-correct attributes of economic theory: that no $20 bills are ever lying in the street—if you think you see one, you must be wrong. Naïve economic theory says that if individuals are allowed to trade and compete in free markets, the resultant efficiencies of production and distribution will be as good as they can get: no $20 bills will be lying on the street waiting to be picked up, and no 30 percent returns on investment in environmental protection will be available when the typical return on investment is 6 percent.

The idea that no $20 bills are on the street is a metaphor for what is known as the "first fundamental welfare theorem" of economics:

"If every relevant good is traded in a market at publicly known prices (i.e., if there is a complete set of markets), and if households and firms act perfectly competitively (i.e., as price takers), then the market outcome is 'Pareto-optimal'; that is, when markets are complete, any competitive equilibrium is necessarily Pareto-optimal." [2]

Pareto-optimal means that no one anywhere in the economy can be made better off without someone else being made worse off. So if markets lead to Pareto-optimality, the economy is doing as well as it possibly can do. It follows that any interference in the market, such as environmental, must come at an economic cost.

A Pareto-optimal society can never have $20 bills lying in the street because if it did, whoever picks one up will make himself better off without making anyone else worse off. (The latter is because the person who lost $20 would not be able to recover it in any event.)

Even in general, the belief that free markets lead to ideal results is an oversimplification at best. And the evidence presented in Part 1 demonstrates that the current economy is overlooking not just one $20 bill lying in the street, but hundreds of billions of them.

Economic fundamentalism makes predictions like the above that are contrary to fact because of errors in its assumption. Many of these assumptions, particularly the more subtle ones, depend on an unstated or taken-for-granted belief that the necessary conditions for free markets can exist without government, or that government involvement in markets undercuts economic freedom rather than protecting it.

The Assumptions of Economics

What are these assumptions? It is difficult to find a complete list of them in economics textbooks: economists tend to forget about what assumptions are needed to support their theories, and instead go straight to the graphs or equations that are the heart of their published articles. Indeed, these assumptions are so overlooked that one must infer them or collect them from a variety of textbooks and studies.

Some of the most fundamental assumptions found in the discussions of economics are:

1. Consumers have practically unlimited material wants, which are self-consistent and rational. All consumers try to maximize their well-being by choosing the optimal set of goods to buy, consistent with their income.
2. Consumer goods are limited, so that tradeoffs must be made between the amounts of consumption of good A with that of goods B, C, D, and so forth.
3. Firms provide goods that compete perfectly with each other, so that no firm has the power to raise prices unilaterally. Similarly, no consumer can affect the price of a good.
4. One person's consumption does not affect another's, so that if consumer X uses more of good A, no other consumers are made better or worse off. So, for example, if I get the chance to consume twice as much ice cream, it makes me happier, and my consumption of the ice cream has no effect on you. Economists express this assumption by saying that there are no "externalities" to my consumption of ice cream.
5. Goods are traded in markets where information is perfectly available; in other words, both buyers and sellers know exactly what they are trading.
6. Each person has a rank order of material desires that enhances his or her well being ("improves his/her utility") and that person A's utility is unaffected by person B:

in other words, if as before I get the chance to eat twice as much ice cream and you see me eat it, your desires for ice cream or anything else remain unaffected.

7. People and corporations always act in their own economic self-interest and behave rationally.

8. Deals or transactions are real—that if one person agrees to sell another person one hundred widgets for $5000, both parties will deliver on their commitments.

9. People and corporations can actually act on their preferences; in other words, if a person wants to buy a particular car for $30,000, she can do so—or if a corporation wants to raise capital for a project with a high return on investment, it can do so.

10. Real market will get us to the optimal point, if it exists, before all of the underlying conditions change.

Each of these assumptions is seriously violated in environmental analyses; often all except assumption 2 are invalid.

What is the significance of this observation about the assumptions necessary for economic theory to work? If the critical assumptions are wrong, then we cannot conclude from theory that government intervention (for example, to protect the environment) will make the economy worse.

Of course, we also cannot conclude from theory that intervention will make the economy better, either.

What we can conclude is that we need to look at the issues on a case-by-case basis. We can also conclude that our theories need to be tested against the data—not unrelated data from other parts of the economy, but relevant data on the environmental issue in question. And when we do so for environmental issues, we see that economic fundamentalism diverts the public discussion away from the real issues. Before we address the ten assumptions above, let's point out a political assumption that underlies the use of economic theory for environmental discussions. This assumption, one that is never stated in the discussions of theory but that rests at the heart of the political uses of economic fundamentalism, is that we have a free market in the first place.

Do We Truly Have an Ideal Free Market?

One of the fundamental pieces of the American dream is that anybody can get rich if he or she is diligent and clever in the marketplace. A Bill Gates can create a company like Microsoft from a standing start through intelligence and application.

Yet the existence of unexploited opportunities like the ability to build a Microsoft, or a Wal-Mart, or to live out any of the hundreds of well-known American success stories, depends precisely on the error in the assertion of naïve economic theory that there are no opportunities that the market has not recognized.

Several reasons explain why this state of affairs could occur, and many of them are important to understand why environmental policies can promote economic growth. First, and perhaps most often forgotten in the political dialogue, is that the current marketplace in America is not an ideal free market. Government-directed changes from the status quo do not necessarily mean a movement away from market forces, but actually could be a movement toward them.

Many of the most common arguments against environmental policy—in particular regulation—take as an implicit but unstated assumption that the status quo is already a free market, and therefore any interference will always produce reductions in economic welfare.

Even if free markets always lead to an optimal result, it would be untrue that to intervene in the real market would take us in the wrong direction. It might be true instead that the intervention enhances market forces and brings the real-world market closer to the ideal.

Now let's look at the assumptions of economic theory and see what are their demonstrable weaknesses.

How Critical Assumptions of Economic Theory Are Violated in Practice

Let's review each of the ten assumptions listed above and see how each is mistaken, particularly for the purpose of environmental policy analysis. We will observe not only how each assumption is invalid, but also how each is wrong in a way that consistently makes environmental protection appear less attractive as an economic proposition than it actually is.

Does Our Consumption Truly Have No External Impact?

The assumption that one person's preferences or consumption does not affect another's (assumption 4, above) is most obviously mistaken in regard to environmental issues.

All of environmental pollution is an externality: all of the consumption that results in pollution clearly affects other people's well-being. The owner of a factory that pollutes, or the customers who buy products whose production involves pollution, imposes monetary and health costs on everyone else (an externality) yet does not pay for them directly. Similarly, an oil company that drills for oil and gas in the middle of a scenic recreation area imposes economic costs on users of the area yet does not pay for them.

The failure of this assumption is widely recognized. Economists believe that this problem of externalities can be addressed if we tax pollution, an approach that is increasingly used in Europe, and is currently actively debated in China. Unfortunately many economists believe that once the pollution externality has been corrected, the problem is solved: no further government interventions to protect the environment are needed. In other words, they think that if polluters have to pay for the costs that their pollution imposes on others, market forces will cause the Pareto-optimal amount of pollution to be emitted—namely, that amount where the costs of additional abatement are higher than the value the incremental abatement would have to those who are exposed to the pollution.

Yet this proposition is obviously incorrect. If the market regularly overlooks two-year paybacks in energy efficiency and a pollution tax raised energy prices 10 percent, the result will turn two-year paybacks into 1.8-year paybacks. Most of these will still be ignored, so the problems go much deeper than merely failing to price the externality.

More broadly, the assumption that one person's consumption

will not affect another is violated in the markets for energy-efficient equipment and designs. If I live in an area where many of my neighbors buy energy-efficient compact fluorescent lamps (which I do), then my neighborhood hardware store will stock a wide variety (which it does)—the products include offerings from at least half-a-dozen manufacturers and are available in a choice of colors, sizes, base types (chandelier base, medium base, and pin base), wattages (from 4 to 45, the equivalent of 15- to 200-watt incandescents), and features (3-way, dimmable, et cetera).[3] So I can easily and inexpensively find a lamp that meets my needs.

Yet if my neighbors ignore compact fluorescent lamps, as is typical in the United States, then I won't find more than one or two choices, at best, in my neighborhood; I will have to travel all over town to find a suitable product, or may fail to find it altogether.

If my neighbors choose to buy organic vegetables, then the neighborhood grocer will stock them, and the vegetables will be easy for me to purchase. If the neighbors don't care, then it will be difficult and expensive to buy organic food.

If the highways are filled with SUVs, the presence of them makes it more difficult for me to read the signs on the highway or even see several cars ahead. The increased risk of death or injury if I get into a wreck with another vehicle is significantly higher with mostly SUVs on the road than if other people drive sedans.

In short, other people's choices have a direct effect on mine, even if they don't affect my preferences.

Yet other persons' choices may also affect my preferences. Many of my acquaintances have expressed dissatisfaction with the option to use compact fluorescent lamps, as they believe the lamps look bad compared to old-fashioned incandescent lamps. They are surprised, however, after they spend a winter evening in my house to discover that all of the lighting is compact fluorescent—to realize they had spent several hours with this technology and not noticed. Such an experience will make them more likely to choose similar lighting. It's evident my preference or consumption does affect how others make decisions.

In the examples above, either person A takes an action that is environmentally preferable and makes it easier for person B to do so, or person A fails to do so and makes it more difficult for B to be green. Therefore the potential benefits to society when person A makes the right choice are more significant than the benefits A receives himself. So person A sees too small an economic incentive to make the right decision, and is therefore less likely to do so.

Do We Truly Have Perfect Information About What We Buy and Sell?

The assumption that we know exactly what we're doing as buyers or sellers (assumption 5 above) is violated almost universally for energy efficiency. Not only do most people lack perfect information about the energy consequences of their actions, in many cases they have no information whatsoever.

Take for example a prospective owner of a commercial building. This investor may not be aware of the amount of the impending utility bills, even when the bills are in the hundreds of thousands of dollars a year. The Energy Star® program and other research efforts revealed, surprisingly, that most building owners haven't a clue about a particular building's energy use.

Appraisers of commercial property actually rely on estimates of energy consumption to help them value buildings when they use the net operating income method of appraisal. In this method, projected revenues (rents) are compared to projected operating costs (these include utility bills), and the difference is called net operating income; the appraisal is based on a multiplier times net operating income. However, the actual numbers for utility costs are almost never used. Instead, the appraiser simply looks up the average for, say, a suburban office building in Chicago, and applies that to every suburban Chicago office building that requires an appraisal. So without information on energy consumption, the commercial building sector (which accounts for some 15 percent of national energy use) can't possibly function as a working market, even in theory.

The situation in regard to accurate information can be even worse for the homeowner. A homeowner or renter has no idea what the energy consumption of any particular energy-using device or practice is, and no way to discover it. Thus a home owner receives no feedback as to whether the struggle to get teenagers not to leave the refrigerator door open has a significant or small effect on energy consumption (it's actually a small one) or how much difference it actually makes to turn out the lights as you leave a room.

All homeowners have is their end-of-month bill.

The above scenario is akin to a trip through a supermarket in which the items are priced by a barcode only and the consumer doesn't know the costs while selecting the merchandise. Certainly, at the checkout stand when the consumer receives the final bill, a total cost for the whole shopping cart cost is provided. Yet because the consumer cannot discern how much each item costs, and because an itemized printout isn't available, the shopper is unable to save money through rational choice (by, for example, the choice of a less costly cut of meat, or the substitution of chicken for steak, or the choice for dried beans rather than meat).

Just as this supermarket shopper will never have the information needed to shop sensibly, energy consumers never have the information needed from their bills that indicates how to use energy sensibly. Unless such a failure is corrected, markets will result in consumers who use more energy than necessary, just as unmarked prices in the supermarket will result in higher grocery bills than necessary.

Such a problem is particularly troublesome for economic decisions that affect the environment. Environmental attributes are almost never traded in markets directly—they are side effects or secondary characteristics of economic decisions on other issues. Because they are side effects, the decision-maker usually will have no information on environmental issues that were traded off.

Here's a simple example—the everyday consumer choice: paper or plastic. The consumer has no idea what are the environmental consequences of the decision. While this is a trivial issue—both choices have minor environmental consequences, and the difference is small—it is illustrative of how larger decisions, too, are made in a way that is inconsistent with the assumption of perfect information.

A more important choice might be whether a high-rise office is built with a concrete frame or with steel beams. The environmental effects of both concrete production and steel manufacturing are significant. Yet neither the architect nor the developer has an idea what they are. A factory owner may have to choose between three alternate production processes that will emit different levels of water pollution. Without government regulations that require limits on pollution or that require disclosure, neither the decision-maker nor his or her customers will have any information about the environmental consequences of the decision.

The problem of imperfect information—in many cases a complete lack of information—introduces a consistent bias into the market. Environmental attributes of products or services will be undervalued because they are unknown. As the assumption of perfect information is violated so thoroughly, markets will always produce more pollution than necessary or optimal.

The invalidity of the assumption of perfect information is an immense problem in today's markets, where pollution is an externality—in which polluters are not taxed—so the consequences of pollution are invisible in the economic system. Yet it would remain a problem even if fees for pollution were assessed. Decision-makers still would not know the environmental consequences—in this case they are also cost consequences—of most of their decisions until it is too late.

Do Personal Preferences Truly Have No Effect On Other People's Choices?

The assumption that my personal preferences don't affect my neighbor's (assumption 6 above) is untrue in a number of ways. First, many goods are positional: they are valued only in relation to other people's goods, rather than on their own merits. Examples include the choice to attend a college based on its prestige, or to purchase a more expensive brand of car that is identical mechanically to another less-expensive brand, or to wear designer clothing that is more expensive than lesser-known brands.

More broadly, one person's preference directly influences another's in many industries. Obvious examples are restaurants, books, or motion pictures. For example, if you see someone at the next table who orders lamb chops and they look good—you are more likely to order them too. And if two friends tell you how much they enjoyed a new book, you are more likely to read it yourself. Consider the fashion industry. This whole industry is built on the fact that one person's preferences do affect others'.

For such cases above, which are widespread throughout the economy, an economic model that assumes that people's choices are independent will give a radically different—and incorrect—answer than one that assumes correctly that one person's choices do, in fact, affect those of others.[4]

The statement that one individual's preferences do not affect another's is patently untrue with respect to location efficiency. If all of your neighbors build smart growth developments yet you continue to live in the same single-family house, your need to drive will go down as a result of your neighbor's behavior. If your neighbors demand and use improved mass transit—so that new service is offered in your neighborhood—that will reduce your need to drive, as well as theirs.

Other energy-using choices can be equally subject to group mentality. If all of your neighbors drive fancy new SUVs, it certainly reduces your own utility to be seen driving something less stylish.

It seems odd that the dependence of my preferences on someone else's should have escaped the attention of proponents of economic fundamentalism as a deep contradiction. Many of the supporters of fundamentalism paraphrase its conclusions with the slogan "greed is good." Their point is that if each person pursues only a sense of individual well-being, or "greed," this leads to the Pareto-optimal result. Yet greed is an emotion closely related to envy—it's difficult to imagine that we could have an economy dominated by greedy people who were not also envious. Envy by its very nature means that one person's preferences are deeply affected by someone else's: one person will always want more than a second person. And if this is true, it undercuts the whole theory that greed leads to the best results.

The failure of this assumption also causes naïve theory to undervalue environmental protection. If others' preferences for environmentally inferior decisions, like to drive an SUV or live in a large, energy-wasteful home—or landscape it in ways that cause pesticide pollution—make me want to choose similarly, either because I like to follow the fashions or to maintain my status, then markets won't provide the Pareto-optimal result. Instead, they will allow too much pollution.

Do People and Corporations Truly Act Rationally to Maximize Their Well-Being?

Assumption 7, that economic units react rationally, is often incorrect both for individuals and for corporations. Its lack of realism is particularly important for corporations. This assumption fails to describe corporate behavior for a reason that economists call "agency." Corporations do not make decisions; individuals who act on behalf of the corporation make decisions. And choices that are in the corporation's interest are not necessarily in the personal interest of the manager who makes the decision.

And even if the manager tries to act in the best interests of the employer, she may lack the information necessary to do so. This is one of the reasons why companies routinely overlook investments with returns of 50 percent per year—these investments may not be in the personal interest of the decision-maker.

Another problem related to agency is that a corporation's paid managers make the decisions; the channels of communication to the owners of the corporation are limited. Certainly a decision about energy efficiency, which typically represents less than 5 percent of the firm's operating costs, would not ordinarily be even communicated from the manager to the owners.

Furthermore, most managers generally are responsible for specific goals for within their departments only. If a manager responsible for capital budget in a building does not also have responsibility for the operating budget, her individual rational outcome differs in a dramatic way from the corporation's rational decision.

Actually the flow of information to a corporation is even more difficult than that. The manager who is responsible for decision-making with respect to energy efficiency or other environmental performance measures probably reports to a vice-president or department manager who in turn reports to a CEO. The CEO gets her information on economic performance through accountants that may work independently of the department managers. Finally the actual decision-making authority of a corporation is vested in the board, which will hear from the organization's CEO though probably not from the department manager, even indirectly.

CEOs have busy jobs, and typical briefing memos from lower-level staff to the CEO must be compressed into a limited number of pages or bullet points. Again, this very structure of communication assures that environmental opportunities are not likely to reach the CEO, much less the board or the owners. They are simply too small in terms of their dollar impact to be included in a 1-page briefing memo.

Such a communication problem can be corrected somewhat if the CEO's compensation is tied to the corporation's performance, and many corporations do just that. Yet this technique still leaves the problem unsolved for at least two reasons. First, the CEO is a human being subject to the limitations of regular people. Recent research has established that people's decision-making departs from rationality in several consistent ways, all of which tend to discourage innovation and risk-taking. (See Chapter 6, for further discussion.) Second, CEOs have a limited tenure of expected employment and so will tend to focus too much on the near term in comparison to the owners, who concentrate more on a longer-term perspective.

A second problem is the difficulty and cost to obtain the information necessary to make rational decisions. Most managers who have the authority over production processes that produce pollution or other adverse environmental consequences, either directly or indirectly, lack the staff expertise to understand what options are available to improve energy efficiency or environmental performance. (See also the discussion of process engineering in Chapter 3.)

It would require a significant investment of time and effort to sort through all the many possibilities to improve economic and environmental performance. And current organizational structures offer insufficient means to do this.

Such failures of the assumption of economic rationality for corporations and other entities managed by agents are well-known in economic policy analysis, but are ignored by economic fundamentalists. They consistently lead in the direction that causes corporations to under-invest in innovations and new technologies that cut emissions and protect the environment.

The problem of agency is only one aspect of the failure of the assumption of economic rationality. Economic rationality is a dubious assumption even for people who act on their own behalf. More than just corporations routinely ignore the opportunities to invest in efficiency or other green practices at a high rate of return: individuals and families do so as well.

The problem of irrational economic behavior by consumers is much broader than this. Clearly anyone who works in the support professions, such as psychiatry or psychology, would recognize the inaccuracy of an assumption of people as rational maximizers of utility, at least for large numbers (a majority?) of people.

Economic theory does not allow for the possibility of someone, for example, who drinks too much on Saturday night and regrets it on Sunday morning; who fails to fill out his taxes on time and has to pay penalties to the IRS as a result of his procrastination; who uses drugs and then becomes sick or disabled—or any other types of activities that ordinary human beings do more than we care to admit.

Hidden Conditions for Economic Theory to Work

Several fundamental but hidden assumptions must be valid for economic theory to work, but these assumptions are so deeply hidden that they are almost never stated by economists or policymakers. Perhaps this is because they relate to legal issues rather than policy issues: things that Americans, with a strong, multi-century tradition of the rule of law take so much for granted that we fail to realize that such rules are actually conscious decisions.

Some rules are listed as assumptions 8 and 9 above, and in a less obvious way, the rules also underlie assumption 5.

1. The rule of law

The most important of these rules is the rule of law and its associated culture of trust. In countries where government is weak or unstable, and where some businesses or individuals regularly act dishonestly, businesses cannot count on the validity of contracts. This undercuts the fundamental basis of markets.

One of the most basic assumptions economists always make, but which needs to be stated explicitly, is the belief that deals are "real." If a business makes a transaction, it needs to count on the fact that the goods or services it contracts for will be delivered on the schedule that is requested and at the place that it's requested, or that they can collect damages from the other company. In the presence of such uncertainty, a marketplace cannot function. I can illustrate the above assumption with an experience I had when I worked with the government of Belarus on energy planning in the early 1990s. One of the fundamental tenets of Belarusian energy policy at the time was to become more independent with respect to energy supply. The Energy Ministry was afraid that supplies of energy from other countries could be cut off arbitrarily, in violation of established agreements, and that Belarus would have no recourse. So the Ministry's goal was to reduce the nation's reliance on energy imports, ideally to zero.

Interestingly, my partners in this delegation included representatives at the California Energy Commission. They noted that California was as dependent on out-of-state energy sources as Belarus, and that the Commission had no concern about this. California energy policy was directed solely toward the procurement of new energy service needs at the minimum cost, regardless of where the energy or energy service equipment was purchased.

The difference was between a functioning free-market economy in the United States—where rule of law assured that California's contracts for out-of-state power would be honored—and a rapidly changing formerly socialist economy in which the rules today could be undercut by tomorrow's decree.

It's challenging to establish a rule of law and its associated social institutions that ensure trust in business deals, so it is unlikely that Belarus was able to change its energy import policy even with the advice of the Energy Commissioner. And it is equally difficult to overcome areas within generally market-based economies where a lack of such assurance that trades that appear reasonable from the viewpoint of economic theory can actually take place in practice. (See the discussion of the California electricity crisis of 2000 in Chapter 5 for an illustration of the problem of how government intervention sometimes is needed to make markets work.)

Deals have to operate under a rule of law and culture of trust in ways that are subject to other conditions as well. If I agree to sell my friend a certain item for a certain cost, both of us know that I actually have to deliver the item, that my friend actually has to pay for it, that the physical transfer of the item is accompanied by a legal recognition of the property right being transferred as well, and that we both understand and agree what the actual characteristics or specifications are of the sold items. To understand what exactly is bought and sold is an issue that revolves around the question of regulation, which is addressed below.

2. The independence of transactions

Another legal assumption that is always made but not addressed in economic analysis of real markets is the potential for threats, intimidation, or special deals. Markets do not always function as theory predicts when one transaction is linked to another. However, special deals are common in real business.

For example, suppose an airline company offers a rebate for a customer who stays at a particular hotel, or offers inexpensive use of their in-flight telephone to a customer who subscribes to a particular cell phone company's services? These linkages undercut the type of competition on which market theory depends.

In the environmental area, one sort of deal overlooked by theory is illustrated when a utility offers an incentive to local home builders to select its heating fuel. Fuel choice has important environmental consequences, but in this case the choice is the builder's—not the consumer's—based on the rebate rather than the cost-effectiveness of the fuel.

The negative deals are more troublesome. A distributor may tell a grocer that it will only supply the one product that the retailer truly wants to sell provided the grocer also stock five other of that distributor's products than those of competitors. Or a manufacturer may only allow its customers to sell its products only if they agree not to sell any competing products.

Even worse, a supplier may tell a purchaser that if the buyer supports a regulation requiring the supplier to act in a specified way (for example, to cut air pollution), the supplier will no longer sell to this company.

Such practices above are widespread in the economy. They may or may not be good business, but they certainly are practices that violate the economist's models of free markets.

And this violation has two adverse consequences for the environment:

First, it consistently makes environmental policies appear more expensive or disruptive than they are, because these special deals make the economy as it actually is depart from Pareto-optimality.

Why does the existence of special deals cause the economy to fall short of the optimum? The answer is because all such special arrangements are designed to prevent a mutually beneficial transaction from occurring. An exclusive arrangement in which a manufacturer only permits a dealer to sell its product if the dealer agrees not to sell a competitor's product is designed to reduce stocking choices. The manufacturer wouldn't need to impose such an agreement if the dealer was unlikely to make more money with increased choices. Should the dealer extend more choices, trades between the dealer and customer would make both better off—the fundamental idea of a free market. Yet special deals in the real market prevent its occurrence.

In some cases special deals impede the marketing of environmentally improved products. Availability of shelf space for compact fluorescent lamps has been identified as a significant barrier to their sales. Similar challenges impede the acceptance of sustainably harvested wood. Due to these challenges, the improved products are more difficult to find and generally more expensive because they must either be sold by smaller or less convenient sales outlets or special-ordered.

The second reason why special deals are a problem for environmental technology is that these deals curtail innovation and restrict competition. Any of the deals described thus far would make it easier for established firms to get new customers and new business compared to new entrants to the marketplace. This is perhaps a small problem in a static economy, but it is a significant problem in a dynamic economy. Special deals make it more difficult for innovative new products to enter the market and attract the attention of consumers, even when the products are superior to those that already exist.

3. Property rights and free markets

The definition of property rights is a key aspect of rule of law that is relevant to the discussion of government's role in environmental policy. Anti-environmental positions often make the implicit assumption that property rights are commonly understood and not debatable, and thus government has nothing to do with them. But in fact, how government defines property rights affects the environmental issue in fundamental ways.

As perhaps the most dramatic illustration of the importance of government definitions of property rights, recall that 150 years ago in the United States slavery was a property right. This property right was removed in 1863, but not without great controversy at the time.

Legal articles have presented how some of the current environmental problems could be described in terms of the absence of property rights for clean air or for a stable climate. If clean air is a property right, then air pollution is a trespass upon that right.

Whether the establishment of property rights in environmental quality is a good idea is debatable, but regardless of one's position, clearly the government's decision to avoid it, whether affirmatively or by default, is a deep-seated intervention in the economy.

A few examples will also illustrate how government definitions of property rights can affect the economic freedom and wealth of an individual subject. First, consider local zoning regulations that limit the type of development that property owners can initiate on their land. Their presence or absence has great economic value. For example, if an urban plot is zoned for three units of housing, its value to a developer is dramatically less than if it is zoned for 150 units.

The number of properties in the United States that are in condominium developments has increased in the past decades. Condominium associations have the ability to impose severe restrictions on property rights, which include limitations on architectural styles that a property owner may use, even down to whether children are allowed, the color of the paint, limitation on the ownership of pets, and restrictions in the uses of the property (such as to prevent the use of an apartment as a place of business).

Some cities have ordinances to establish property rights in the view from hillside homes. In such cases, the homeowner who allows a tree to grow on the property in a way that impinges on the view of the neighbor uphill has the obligation to trim or remove the tree. This is similar to an environmental-protection property right because, by extension, one could argue that the owner of a property with a view could have a similar property right in the assurance the view remains unsullied by smog and soot from air pollution.

The definition of what property rights truly are makes a critical difference to environmental policy. For example, I live in a hillside neighborhood in San Francisco where many of the homes have dramatic views. A preferred view will raise a property's value by 40 to 60 percent.

However, such views will only have this effect if air pollution continues to be low. If I can see Mount Diablo, which is forty-five miles away, only one or two days a year, the view will be devalued. I will suffer real economic damage if I have to sell and air pollution has degraded the view. So arguably I have a property right in clean air.

The right to clean air is not recognized in the American legal system today—but it could be, in theory.

The broad point is that the choices the government makes in the definition of property rights affects the evaluation of whether a particular set of market institutions leads to an optimal result. "The fundamental theorems of welfare economics can only be true with respect to a particular structure of property rights. And then it is no longer possible to claim that any particular competitive equilibrium is Pareto efficient. A different property rights regime could—and probably would—make everyone better off."

The foregoing discussion is not a reason to assign property rights to the environment, but rather to demonstrate two critical problems with economic fundamentalism. First, it shows that the belief that government policy to protect the environment will cause the economy to depart from Pareto-optimality is wrong. The answer will depend on the circumstance of how the government defines property rights.

And second, the discussion illustrates just how deep-seated is the role of government in the economy/environment discussion. The government cannot just leave environmental protection to the market and hope for the best. The government, by action and by omission, defines the market and produces different results from perfect competition, which depends on what decisions it makes.

More Hidden (and Inaccurate) Assumptions in Economic Fundamentalism

So far the discussion has focused on explicit assumptions that are necessary for the principles of neoclassical microeconomics to work. Some hidden assumptions about how the economy works must also be valid to support the theory that markets lead to the best outcome.

As we've seen above, Pareto-optimal means that nobody can be made better off without someone else being made worse off. This is the best possible outcome in economics and it is understandable why economists rely so heavily on this concept.

Economists refer to Pareto-optimal as an equilibrium because theory says that once the economy is at the optimal point, it will never move elsewhere. But will it ever get there in the first place?

Economic fundamentalists often assume that if competitive forces operate fully in an economy, the competitive equilibrium will be reached naturally through the dynamics of the marketplace. This process can be analogized to a steel ball that rolls on an undulating surface: the equilibrium point is the lowest point on the surface, and the ball will always roll downward until it reaches the low point.

In fact, the approach to economic optimization is not well understood. "The lack of a well defined equilibrium finding mechanism has been an embarrassment since the earliest days of mathematical economics."[6]

In contrast to the ball, which gravity pulls down and friction slows down until it comes to rest at the low point of the surface, the process to reach the equilibrium point—the Pareto-optimal point—is unexplained in economic theory.

Nothing requires a firm in a competitive market to take advantage of another step toward the equilibrium point unless that step is necessary to keep the firm in business. A firm might miss all sorts of opportunities to increase its profit, but as long as it remains profitable overall, it can stay in business and continue along its bumbling path.

The approach to optimization is a process akin to how you feel your way through a maze in the dark. You cannot see the shape of the whole maze. At best, you can tell if you are at a dead end or if you are failing to approach the goal. In the context of an economy in which anything changes, even at a modest pace, a slow approach process toward the optimum guarantees that the economy will never get there.

This problem is even worse in a dynamic economy. If technology can advance, then the optimum point shifts over time. Yet if it takes too long to reach the optimum point , that point will have shifted before the economy can come close to getting there!

To reach equilibrium is an even more challenging problem than that. An economic optimum means the best point for the entire economy; an analogy might be to find the lowest point on the surface where the ball rolls. If only one low point exists, a simple rule of downward movement at every step will take you to the bottom of the surface. A rolling ball will always reach the lowest point

However, if local optima are present, as well as global optima—that is, if there are shallower low points on the surface in addition to the deepest low point—a ball that starts to roll slowly downhill near a shallow low point can only get to that nearby low point and has no way to climb again and find the lowest point. Indeed, given the difficulty and cost to acquire information about where the optimum lies, even for an individual firm, observers of and participants in the economy may not even know that the economy has missed the true optimum. Metaphorically, if the ball rolls on the surface in a fog and falls into a minor low point, rather than the lowest point on the surface, the viewer may not even see that it is stuck.

To compound this problem further, observation and supporting calculations indicate that often an economy has more than one equilibrium or optimum, and thus the point that is most optimal from one point of view or any particular point in time may not be the only such solution.[7] Thus, a change in an ideal market is not necessarily a change for the worse, particularly in a dynamic and changing economy, but even in a static economy.

The Need for Regulation

If we look carefully at the first fundamental welfare theorem of economics, we discover additional subtleties that are relevant to the question of why markets fail to produce economically justified investments in energy efficiency.

The theorem begins with the statement, "If every relevant good..." This definition assumes that "relevant good" is understood. But this is not usually the case.

Illustrations of perfect markets in economics textbooks often focus on trade of simple commodities like corn or gold with the unwitting assumption that everyone knows what a bushel of corn or an ounce of gold is. Yet even for these simple commodities, regulations are needed to define the product. Corn harvested this year is not the same as corn that has been stored for ten years. Corn produced with genetically modified plants is not the same, at least to some traders, as conventional corn. Corn with a high level of contamination by vermin or pesticides is not the same as cleaner corn. And even for gold we need regulations to define purity and accuracy of weight.

Defining Terms

In the real business world, if an investor buys a contract for a thousand barrels of oil on the commodities market, both the buyer and the seller must understand a lot of legal and technical conditions that must be met for this transaction to take place. The legal conditions refer to the time and place of delivery of the oil. The technical conditions, which no one thinks about in most cases because they are so well addressed by the current legal and regulatory system, are the definitions of the commodity itself. What exactly is a barrel of oil? What are the chemical characteristics of the oil? What physical characteristics does it possess, and what characteristics (e.g., contaminants) are absent from it?

To assure that all parties understand what exactly they contract for requires a rather extraordinary degree of regulation and standardization. Government does some of this regulation, and the private sector also does much of it. Some is done informally, and some regulation is accomplished by gentlemen's agreements that are not even at the level of informal regulations. Yet without regulation at some level, commerce simply cannot exist on a rational basis.

Imagine two companies that arrange to sell a ton of steel but disagree on how to measure the weight of the steel, what the content of the steel is in terms of different metals, what its strength is, whether it's rusted or not, et cetera. Clearly this problem becomes more severe the more complicated and high tech the item of trade is.

However, the problem is not limited to high-tech items in industry. For example, consider a used refrigerator as one example of the problem of defining the relevant good. The potential buyer of the refrigerator can tell what the manufacturer, size, condition, and likely performance of the refrigerator is. Yet the buyer has no information whatsoever on the amount of energy bills the product will generate. And the lifetime value of these bills may exceed the purchase price!

This is true almost universally for energy-related purchases, except to the extent that the government has intervened in the market and required that products or buildings or industrial processes bear labels that rate them for energy efficiency. So if we can't even define "relevant good" in a consistent way without government policy that provides standardized test methods and assures the availability of information, the most fundamental assumption of welfare economics is undercut.

Understanding "Relevant Goods"

Many economists, and virtually all economic fundamentalists, fail to recognize the extra- economic process that is necessary to define what "relevant goods" truly are. We all think we know what we mean when we describe a transaction for a house, or a ton of steel, or a gallon of gasoline, or a refrigerator. But in fact, numerous standards and criteria exist, some explicit and some implicit, that must be assumed by both seller and buyer in order to make a transaction work, as well as in order to make it consistent with economic theory.

The need for regulations to describe the qualities of goods in the marketplace has been in evidence for thousands of years: the Bible documents repeatedly the moral imperative to use honest weights and measures, and in some cases it actually specifies exactly what they should be, and to sell exactly what you purport to sell in market transactions.[8]

Note that this discussion refers to regulation of quality and not price. Price regulation is an important part of the polical/economic discussion on a number of issues, which includes energy issues. But it is not part of a discussion of why economic fundamentalism relies on false assumptions.

Regulations begin with the descriptive basics of physical units and time. What is a pound? What is a foot? What is a second? How accurately do you have to know them for any particular application? If a bag of potato chips lists its weight as one pound, but the weight is actually .999 pounds, does that constitute legal fraud? Where is such a line drawn?

Regulatory needs of a market economy then proceed to numerous definitions of quality, functionality, safety, comfort, and compatibility of a particular product with other products. The computer concept "plug and play" is an example of regulations based on a complex series of requirements that concern both the peripheral device and the computer.

Standards describe such things as if your television receiver will accept a broadcast or cable signal, whether the output of your software is readable by someone else's, whether the replacement keys from your car can be ground from blanks at the neighborhood locksmith, and so forth. In fact, the more one thinks about the areas where regulations are needed and used, but invisible, the longer the list grows.

Regulations Encourage Competition

Regulations are a necessary condition to free-market competition's working at all. Economic theory casually introduces perfect free markets that allow competition on the production of a given good at a given price. In order for competition to exist at all, different firms must produce the same product. Without regulations, no one can know whether different firms' products are actually the same!

Without regulations to specify exactly what a product is, a company that wants to purchase a product that embodies a particular attribute that is needed would have to specify a particular brand name and model number. Such a specification eliminates competition entirely: only one company could respond to a bid.

Without regulations, all markets would tend to be monopolies at worst, or oligopolies (where only a limited number of firms compete) at best. In markets without regulations, buyers could at most trust the limited number of suppliers that they could check up on to produce the quality of product that they chose to buy. Decisions would have to be made not on price and quality, as economics calls for, but on reputation and trust. With regulations, a limitless number of companies can compete, and compete on a level playing field.

Setting Standards

Many regulations are established in the private sector. Standard-setting in the private sector has not been addressed extensively in the economics literature.[9] Yet even in 1983, there were 32,000 formal private-sector (which includes the nonprofit sector) standards that were maintained in the United States by some 420 different organizations. These private-sector standards have been in existence in some cases since before 1840.

By 1991 the U.S. National Institute of Standards and Technology estimated that the number of private-sector regulations had grown to 41,500.[10]

Clearly these standards were set in place at considerable expense, because business needed them. And business willingly pays for them.

For example, the Air Conditioning and Refrigeration Institute, the trade association for the air-conditioner industry, employs almost half of its staff on the development and certification of industry standards. This effort is supported financially by that association's membership, which indicates that the regulated companies find it very important to have these regulations in place and up to date and enforced. Interestingly, most of these companies would not even consider themselves to be regulated but would instead believe that they were operating solely by the rules of market competition. And this belief would be correct—because regulations underlie competition.

A lack of regulation can impede the development of industry. For example, in recent years substantial debate among different companies in the United States has occurred about how to commercialize high-definition television (HDTV). Different companies had different proposals for standards, and the failure to agree on a common standard repeatedly set the industry back. Finally the government imposed a decision and established a date when televisions would have to be compatible with HDTV. This was broadly perceived as a pro-business step that allowed industries to make the investments necessary to commercialize a new and potentially profitable technology. Without a common standard, this could not have gone forward.

One of the asserted reasons for the greater use of cellular telephones in Europe than in the United States is the European Union's faster move to establish a common standard for cellular communications (the GSM standard) while the United States developed four mutually incompatible systems of standardization.

Thus far the discussion has been limited to formal written standards that can be referenced. Yet these are supplemented by informal standards that are used in many industries that are not actually written as standards, but which guide business practices in a similar manner.

Theory Versus Reality of Free Choice in Markets

Informal private-sector regulations can severely limit consumers' and business's free choice, in contradiction of assumption 9 above that everyone can act on their economic preferences—that a consumer can choose to buy as much or as little of a good or service as desired. For example, if a consumer wants to buy a $30,000 car, she must either have the cash to pay for it or the credit to finance it. However, the availability of credit is determined by informal private-sector regulations, and some potential buyers may not qualify.

The situation is even worse for the housing industry wherein informal but rigid lending regulations specify how mortgage lenders decide to make a loan or not, and also decide what interest rate to charge. These rules in turn often cite privately compiled credit scores, which act as a standard even though not formally written as such.

Commercial real-estate development also has numerous standard practices in regard to what types of projects are considered low risk and easy to finance. Smart growth developments do not fit these regulations because they are different than the more familiar sprawl developments, and the ability for the developer to get the project financed is a major problem for many such projects.

Businesses that apply for credit are subject to parallel informal standards about how they must act and what paperwork has to be filed in order to qualify. These standards often affect business decisions in a fundamental way, not just peripherally. For example, credit ratings of corporations can limit whether they can undertake large projects. Credit ratings can also affect the economic options of whole countries.

Informal regulations on credit-worthiness are one of the important barriers to energy efficiency. Suppose a building owner has identified a $1 million investment in energy efficiency that will save $400,000 a year. However, if the owner is a corporation whose current debt is too high based on the informal regulations, the owner will not be able to raise the capital for the investment. Worse yet, even if someone agrees to loan the owner the money (the local utility, for example), these regulations may still prevent the company's acceptance of the loan. Economic theory as propounded by the fundamentalists would assert that a corporation can make whatever decisions it finds profitable; but a realistic theory must admit that other actors may control the choice. This problem had a deep consequence in the failure of California's experience with so-called free markets in electricity, which is discussed in the next chapter.

In the case of lending rules, both for developers of smart growth projects and for their prospective buyers—who may want to spend less on transportation but more on housing than is traditional—regulations are what prevents environmentally responsible actions, not what causes them.

So to frame the environmental question as regulation versus market forces is fundamentally inaccurate. Regulations are a part of market forces, not an alternative to them. And regulations can enhance market competition and consumer (and business) choice or they can restrict them.

How Markets Actually Work

All of the considerations thus far point in a single direction: they undermine the argument by many advocates of the status quo—that markets, which are assumed (incorrectly) to operate free of government interference, provide the best economic answer for everyone, and that additional government involvement, be it by regulation or by a change in the incentives or the rules or terms of competition, must therefore decrease economic welfare.

Instead, this analysis of the assumptions that underpin the theory of market forces demonstrates the opposite: for markets to function as closely as practical to the ideal, a strong, continuous, and highly interventionist government is necessary.[11]

This observation about the important role of government intervention in facilitating free markets and competition undercuts several myths that have influenced popular thought on the relationship between government and the planned economies. One such myth is the belief that the failures of Soviet-style economics were due to the fact that the government controlled everything. The parallel myth then is that the successes of capitalism are a function of the extent to which the opposite is true, mainly that the government controls nothing. Actually, the government didn't truly control everything in Russia during the Communist regime: it made the high-level decisions about output levels, input supplies, and prices, but it had a difficult

time getting high-level decisions to be carried out at the level
of actual production.

An analysis of the assumptions that need to be satisfied for
markets to work shows that the proposition that markets work
best to deliver growth and economic well-being is not proven
or disproven from general theory. Instead, a more detailed look
at the specific issue in question—a look that studies the data as
well as the theory—is needed.

In general, markets work best under the following conditions:

- Governments and social institutions are stable and strongly
  enforce consistent rules of law and consistent systems that
  determine property rights.

- Informal attitudes and regulations reinforce government-
  sponsored rule of law.

Financial markets depend on government agencies and cul-
tural attitudes to reinforce levels of trust that assure that if some
of the myriad details of what a transaction means and who is
responsible for what are not identified more clearly, that reasonable assumptions can be made to fill in the gaps in a way
that is fair.

The failures of economic fundamentalism do not undercut
the belief that markets work better than such alternatives as
government central planning. The fact that the real economy
fails to produce the Pareto-optimum result does not mean that
economic welfare is unlikely to be maximized by the natural
operation of ideal markets. It means that economic principles
need to be applied to the details of how real markets function.

So if particular assumptions that are necessary for markets to
produce optimal results are violated, we need to look at the facts
of how the assumptions fail and what their consequences are.
We then should look at new interventions, or changes to current interventions, or to elimination of current interventions,
that will cause markets to function in reality as they are believed
to function in theory.

On environmental questions, it is possible to identify in detail many of the areas where markets are successful and where they fail. And when they fail, we can often describe the mechanism by which the failures occur, and design policy interventions that correct the failures or compensate for them. And even when we can't describe usefully how markets fail, we may be able to identify corrective policies to improve them.

The issues of how markets can fail are also discussed in Chapter 6. I will illustrate how an identified set of failures of the market systematically causes an under investment in innovation and new technologies. When these technologies affect the environment, these failures lead to much more pollution than is economically defensible. And if we correct this under investment we will advance growth, prosperity, and job creation, as well as achieve the direct goal to enhance environmental quality.

A surprisingly large fraction of self-described conservatives' arguments fall into a trap that unwittingly assumes markets mean economic fundamentalism. They implicitly assume that markets with the minimum of government involvement give the best economic result and that changes that rely on government always make things worse.

Such are the attitudes that contributed strongly to the failed California experiment in electricity restructuring, which is described in the following chapter.

5 | Lessons from California's Failed Experiment in "Free Markets" for Electricity

In 1992, the California Public Utilities Commission (CPUC) initiated a discussion of how the state's utility system could be "deregulated" to rely less on the institution of franchised monopoly utilities and more on market forces and competition. This discussion culminated in legislation that set up hourly markets in wholesale electricity, wherein price was determined by supply and demand. Yet soon after this system was implemented, the new regulatory regime failed dramatically in the year 2000.

The Road to Failure

The California Energy Commission identified the first steps on the road to failure some eighteen months before the crisis hit. The Commission noted that no new power plants had been built in the preceding years—at the same time when the utilities were slashing investments in energy efficiency (in response to the new regulatory system, as described below). Demand continued to grow at an annual rate of just over 1 percent.

With several years of demand growth, no new generation, and diminished expenditures on energy efficiency, the Commission predicted a high likelihood of power shortages as early as the year 2000, and began to urge the state to take action. However, these pleas were not attended to promptly, and actual shortages began to appear by summer 2000.

By late spring 2000, wholesale electricity prices skyrocketed almost tenfold, which cost the state $15 billion in extra electricity costs—about $1,000 a household—and sent the state's largest utilities to (or over) the brink of bankruptcy. And the lights nearly went out for everyone.

As serious as the 2000 experience was, 2001 was predicted to be several times worse. But an emergency program of efficiency and conservation averted the crisis.

The failed California experiment in the 1990s with electricity restructuring demonstrates both the role of economic fundamentalists in setting up a system that ignored a century of experience with the utility industry and imposed the simple ideology of the perfect marketplace onto a complex system and the prevalence of the economics-as-religion paradigm in an attempt to explain what happened when the newly constructed market design was subjected to different market conditions than the regulators had anticipated. Restructuring resulted in an economic meltdown for utilities and electricity consumers in 2000, only two years after the new and deeply flawed market design was initially implemented.

Bankruptcy, Credit Failure, and Higher Prices

The simple facts are that in the year 2000, shortages of electric power in California led to increases in wholesale electric prices from the normal range of about $0.04 to .06/kWh to the remarkable range of $0.20 to $1.00/kWh and higher.[1] This resulted in some $15 billion of additional unexpected costs for electricity that year in California alone.

Because retail rates were for the most part frozen by regulation, California's utilities absorbed the cost increase at first. CPUC regulators made certain Southern California Edison Company (SCE) and Pacific Gas and Electric Company (PG&E) did not pass along the increased costs to their customers, although both were required to continue making mandatory power purchases in the wholesale electricity market. Ultimately they were financially unable to handle an increase of such magnitude. Thus PG&E, the largest utility in the country, filed for bankruptcy in April 2001. SCE, one of the next largest utilities, teetered on the brink.

Worse yet, the deteriorating credit-worthiness of the utilities threatened even a greater disaster: electricity suppliers began to state that after some specified date in early 2001, they would no longer sell electricity to California at all. The risk of intermittent rolling blackouts had grown to a threat of a continued statewide blackout.

Eventually the state government—which at that time had sufficient credit-worthiness to assure sellers that they would be paid—stepped in to guarantee the availability of electricity through purchases of long-term contracts to supply electricity.[2] However, due to the continued power shortages, the contracts committed the state to relatively high prices.

The electricity problem was predicted to get much, much worse in 2001. The excess cost of electricity was predicted to be $40 billion in 2001—or well over $3,000 per household—a dramatic increase over the $15 billion paid in 2000. Some forty days of rolling blackouts—days when some areas would be without power in order to protect the entire electrical system from failure—was predicted.

Yet this time the state took aggressive action. California invested $1 billion in emergency appropriations to promote a crash program of energy conservation and energy efficiency, which completely solved the problem: no rolling blackouts were recorded in 2001, and electricity costs returned to a normal level.[3]

Shortly after the problem hit, a number of mythical and incorrect explanations of the causes of the crisis were circulated. The refutation of these myths, and the detailed descriptions of how restructuring began and how it was followed, provide interesting insights into the role of economic ideologues in decisions that set environmental/economic policy.

What Actually Happened—Myth Versus Reality

The myths of economic fundamentalism, as advocated in the 1990s by both public officials and by powerful industrial stakeholders who were concerned about their cost of electricity, caused regulators to make radical changes in the electricity market.

Before this, utilities had been vertically integrated franchised monopolies, with prices (electric rates or tariffs) set by state regulation. Utilities owned the wires and distribution stations that transferred electricity from the power plant to the customer's home or business, and they also owned the power plants. Utilities were allowed to price electricity at a level that would pay them back the actual costs they incurred as well as give them a reasonable rate of return on investment.

Utilities were also charged with the provision of reliable electric service: they were required to forecast how much power would be demanded and to respond with sufficient quantities of power and transmission to meet foreseeable needs with a margin of safety. This forecasting and planning process gave utilities the ability to meet needs through efficiency as well as new power plants and transmission lines. The whole system was subject to state regulatory oversight.

Everyone fairly well accepts the wisdom of a system of monopoly provision of wires: it would be wasteful to allow competing companies to install separate sets of wires to each customer.

However, generating power is different: building power plants can be a competitive business, and many policymakers were headed in a direction that would allow competition in the generation of power. Indeed, California was in the forefront of taking the monopoly out of generation: as of the early 1980s the state required utilities to buy power from any reputable seller at a price fixed by the Public Utilities Commission. By the early 1990s, this policy led the state to have the highest level of renewable electricity generation in the world.

But, the state requirement for independent power producers to be able to compete with utility-owned power was not enough for the economic fundamentalists. In the early 1990s, they began to agitate for what they believed was a completely competitive system in which any buyer could make a contract with any seller to buy electricity at a negotiated price, and transmit it over the utilities' wires.

The planning process, by which utilities forecast needs and provide sufficient capacity, was to be replaced by a market-based system, in which only nonutility companies would build power plants. Market forces would determine sufficiency of supply: if supplies appeared tight, the prospect of higher prices would induce investment in power plants.

The idea of individual contracts between users of electricity and generators was extremely attractive to large industrial users of electricity who were confident that they could negotiate lower rates this way.

Economic fundamentalists loved the approach because it allowed markets to replace regulation and planning.

This combination of large and politically powerful incumbent industries and economic fundamentalists proved impossible to resist, and in 1992 the Public Utilities Commission set out on a path to deregulate the electric industry.

The momentum for what its proponents called deregulation (or, as those less enamored of the idea called it, "restructuring") was so strong that policymakers ignored crucial reasons why market forces could not be expected to deliver the right answer.

As a result, in 1998 regulators established a new electricity marketplace. This market was soon found to have major conceptual as well as design flaws (most of which were predicted by critics of the headlong rush to restructure) that did not manifest themselves until the state was confronted by physical shortages and skyrocketing electricity costs in 2000. Both utility regulators and other state agencies delayed their response to these problems until 2001. Taken together these factors led to a significant amount of damage.

Even after the damage was done, economic fundamentalist advocates would not admit the failures of their plan, and began to explain the disaster in terms of myths.

The main myths:

Myth 1: The problem of an energy shortage was caused by large unexpected growth in consumer demand for electricity, primarily powered by the success of the electronics/Internet industry.

Myth 2: The problem was that no new power plants had been built in California due to one or both of the following problems: environmentalist opposition and approval for power plants was a regulatory nightmare

Myth 3: Greedy utilities lobbied for and obtained passage of the restructuring bill, Assembly Bill 1890, in order to pass on to customers the high costs of their mistaken investments in power plants.

Myth 4: Anyone could see that the market structure set forth under the restructuring rules was flawed and would lead to disastrous results, yet special interests (utilities?) allowed it to go through regardless.

Myth 5: The principal problem with restructuring was that it didn't go far enough: deregulated costs should have been passed through to consumers rather than absorbed by utilities. Higher consumer prices would have led to reductions in demand and increases in supply that would have in turn prevented (or at least greatly mitigated) the problem.

Myth 6: State regulators, without regard to the consequences, restructured the markets to rely exclusively on short-term, spot-market sales of wholesale electricity, an obvious mistake. (Of course, this was not the entire explanation for the crisis, but it was an important contributor to the flawed market design and the imbalance in supply and demand that resulted.)

While some extremely limited truth lies behind the last three of these explanations, the facts bluntly contradict Myths 1 through 3. A brief discussion of what really happened is presented below. (Appendix A provides a detailed refutation of the myths.)

In summary, here are the facts:

Reality 1: California's electricity use grew at a relatively constant rate throughout the 1990s—a rate that was correctly forecasted by utilities and their regulators and was lower than that of the rest of the United States.

Reality 2: Environmentalists supported new power plant construction because the new plants either used renewable energy or burned natural gas with high levels of pollution control. State licensing procedures had been streamlined in the 1970s, and in the early 1990s, the California Energy Commission certified eleven power plants for construction, eight of which (totaling 960 MW) were ultimately completed. But not one power plant application reached the Commission from 1994-1997. The reason no plants were proposed after 1994 is that investors did not want to put money into them given several years of uncertainty over market structure. Although large out-of-state companies bought California's older natural gas-fired electric-generating plants in 1997 and 1998, by the time of the energy crisis they had not brought new or more efficient units on line. It was the lack of investor interest in power plants during a period of perceived power surplus exacerbated by cutback in efficiency programs and an abrupt end to the purchase of renewable power sources that caused the actual shortfall.

Reality 3: Utilities initially opposed restructuring, which was the economically rational position for them. Restructuring was promoted by big users of electricity who wanted lower rates, and by economic fundamentalists who disliked rate regulation that they believed subsidized residential rates at the expense of industrial rates (or who thought they could make money by designing a complex new system of electricity markets). Utilities stood to lose the battle, however, and, along with many other stakeholder groups, accepted the state restructuring law because it gave them a better deal than what the Public Utilities Commission otherwise had offered.

Reality 4: A team of consultants who insisted that they had carefully studied utility systems worldwide developed the system design adopted for California's new electricity market structure. They claimed the system was tailor-made to be the most competitive system possible for California. Both the Public Utilities Commission and legislators dismissed concerns over the market structure.

Reality 5: Passing the costs on to consumers would not have affected usage (or price) much in the short run (that is, from Spring 2000–Summer 2001, the period of the crisis) because the first rise in costs—a doubling of retail prices from one month to another—was passed on to San Diego consumers and the decline in usage was only 2 percent. To pass on the full costs to consumers was a technical impossibility, at least in the short term because wholesale electricity prices fluctuate from hour to hour but cost-effective home electric meters that adjusted prices by time of day were not available.

Reality 6: The reliance on spot markets was intentional. The restructuring program's designers and the regulators at the Public Utilities Commission believed that the best way to accomplish long-term goals was through financial markets rather than regulatory-directed long-term contracts. The PUC insisted that the price transparency of short-term spot markets was essential and it relied on the "genius of the market" to develop inter-mediation contracts that would turn short-term transactions into long-term ones, just as financial markets allow thirty-year mortgages to be supported by bank deposits that can be withdrawn at a moment's notice. The "free market" theory that was the whole reason for restructuring electricity markets in the first place predicted that new financial markets would solve this problem better than government policies or deal making by monopoly utilities. After 1995 the Federal Energy Regulatory Commission, later joined by the PUC, began to refuse approval of long-term contracts for power, so utilities did not renew them. Thus by 2000–01 out-of-state supplies previously committed to California were free to pick better-priced markets and to satisfy their own local, native load that had grown more rapidly than California's.

The focus on myths allowed the economic fundamentalists to divert the discussion from the flaws in their plan to reasons why other people were responsible for the failures.

So let's first look at the errors in the economic fundamentalists' worldview, errors that should have been apparent from the beginning of the debate in 1990.

The key error was to put the cart before the horse: to focus on the answer ("market forces and competition") before the question ("what is wrong with the current system of electricity production and consumption?").

This error was so critical because the only answer to the question of what is wrong with the system is that the prices were too high. Proponents articulated no other reason for deregulation. And the reasons that prices were too high were well-known: utilities had made long-term commitments, in many cases in response to state regulation, to sources of electricity that turned out to be more expensive than new gas-fired power plants.

The only way that markets could solve this problem of high-priced power plants and contracts would be to force utilities to write off their long-term commitments at a loss of billions of dollars. If such write-offs were politically desired (which did not turn out to be the case), this could have been done directly: deregulation was not the only means to accomplish this objective.

One could have argued that a competitive market would also allow new power plants to be constructed and operated more inexpensively. Yet no one made this argument, or at least no one tried to estimate what the cost savings from this new element of competition would be. Some proponents argued that competition would drive electricity prices down, but they could not offer a reason why these prices should be lower than those produced through utility contracting.

Indeed, none of the proponents of deregulation was willing to admit that if market forces rather than utility planning were to determine whether and when power plants would be constructed, one possible outcome could be (and in fact was) that periodic shortages would occur and prices for electricity could temporarily become remarkably high. From that point of view, a self-consistent economic fundamentalist would have to argue that the high prices California saw in 2000 were all right: a result of the natural function of the market.[4]

So the only real arguments for restructuring were that the

state could create new markets for the sheer joy of it, without any evidence or even calculations of how much money (if any) would be saved, and the business advocacy of large industrial users who thought that the system would lower their electric rates.

The weakness of these arguments is now evident—rather than save money, the restructuring experiment cost California over $25 billion ($15 billion in overcharges for electricity, $1 billion for an emergency conservation and efficiency program to mitigate the damages the next year, and some $10 billion in extra costs for the long-term contracts negotiated in early 2001) and also imposed large costs on neighboring states.

Cuts in electricity rates never occurred. Rather, restructuring raised rates by 40 percent. So it follows that economic fundamentalists would want to distract the public from this type of discussion and instead focus on the myths.

The Consequences of the Restructuring Experiment

The demand for electricity in California grew slowly but steadily during and after the debate on restructuring, yet less slowly than it would have, because of cutbacks in efficiency programs that resulted as a response to restructuring. Supply stagnated due to uncertainties among investors coupled with a market perception of significant power surpluses.

In truth the imbalance between demand and supply was much worse because California is part of the Western Grid—not an island. The Western Grid embraces the eleven western states. California represents only 40 percent of total power on the grid. The other states accounted for 85 percent of the growth in power demand on the grid since 1995, so the stress represented by California's growth was greatly magnified by growth elsewhere.

The Rise in Prices

The crisis struck first during the summer of 2000 in San Diego where, due to a quirk in the restructuring law, retail prices were deregulated a year earlier than elsewhere. As wholesale prices skyrocketed, San Diego area residents' electric bills doubled in a single month. This generated considerable outrage: the chairman of the California Senate Energy Committee and the architect of the restructuring law suggested in public that consumers simply not pay their electric bills.

Consumption of electricity in San Diego declined 2 percent in response to the two-fold increase in prices. Political outrage by consumer and small business organizations rescinded the increase in prices at the retail level, despite a continuation of high wholesale prices. When the crisis hit, California prices started to follow the pattern illustrated by the curve in Figure 10 below.

Figure 10
The Cost of Electricity at Peak Demand

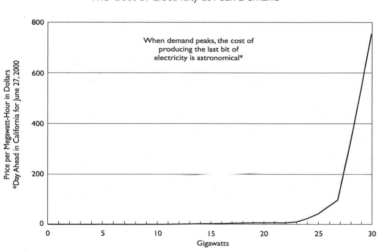

Source: *Business Week*, March 26, 2001

Figure 10 illustrates that as long as demand remained below a certain threshold (about 23 gigawatts), prices would not differ significantly from historical experience. Yet once that threshold was reached, prices could skyrocket. California spent virtually the whole summer riding up and down the steep slope of the so-called "hockey stick."

Recently many allegations have surfaced over energy supply companies' manipulation of the market. In their most extreme form, these concerns could be considered yet another myth: that the shortage was phony—contrived by energy suppliers in order to squeeze consumers. This is not a myth of the economic fundamentalists—it is inconsistent with their world view that market manipulation can exist—rather it is a myth of the left. But the responses to Myths 1-6 above also refute this belief: clearly electricity use was increasing while new power supplies were being deferred. Clearly the impetus for restructuring was coming from industrial power users and economic fundamentalists and not from the generating companies.

What is important from a policy perspective is not the extent to which suppliers manipulated the market. The point is that the structure of the markets and the prospects of physical shortage guaranteed that the conditions for vulnerability to market manipulation would exist in California. The very shape of the supply curve in Figure 10 by itself constitutes a red carpet rolled out to any owner of a power plant to act in socially nonproductive ways. The slope of the curve is so steep that if a single owner had two power plants of 400 megawatts each, the decision to run the plants or shut them down would have a noticeable effect on price when demand reached certain well-known, foreseeable levels.

If the slope is steep enough, which apparently it was, a power plant owner could make more if he sold 400 megawatts of power at more than double the price than he could with 800 megawatts at the price that would have resulted if all 800 megawatts were placed on the market. Even without criminal manipulation of the market, the very bidding structure when applied to the current inventory of plants assures that markets will be manipulated to the public's disadvantage because everyone gets the market-clearing price of the last accepted bid.

And even if regulators could have stopped the exploitive

deal-making that occurred, California still would have had a big problem with shortages and high prices.

Note how the shape of the curve violates one of the most fundamental conditions of how a competitive market is predicted to function: that individual corporations must be price takers—that no individual producer or consumer can affect price. However, with a steep curve such as that in Figure 10, virtually any individual player can influence price. Because in California only a finite number of power plants exist, whose characteristics are well-known, this outcome could have been predicted if market conditions similar to 2000–01 had been analyzed in advance.

More broadly, the error of the economic fundamentalists was to forget that for an ideal market to work, there must be no externalities: one person's consumption must not affect anyone else's. But electricity has the opposite characteristic: an electric supply system is all interconnected, which means the entire system of power plants, transmission lines, transformers, and millions of users amounts to one giant machine. Everybody's consumption and production affects everyone else's, especially when supplies are tight.

Electricity can't be shifted from one transmission line to another the way soybeans can be shipped on one railroad car or another, electricity rolls "downhill" from higher voltage (at power plants) to lower voltage (at centers of consumption) over all existing wires. Regardless of who contracts to sell electricity and who buys it, and regardless of who owns the wires, the flow of electricity is the same. And if the amount of power generated at any second is not equal to the amount consumed, the system will black out.

This is exactly what we saw in 2000. Everything was connected to everything else. One consumer's shortage was all consumers' shortage. One power plant's shutdown raised prices for everyone (which included the inoperative plant's owner). As high prices continued throughout the fall, and even into winter of 2001, the utilities' financial situation became untenable. Not only were the utilities on the verge of bankruptcy, but they were also so weak financially that power suppliers threatened not to sell them electricity at all, because of concerns of nonpayment.

The Risk of Extended Blackout

California was at risk of a prolonged, total blackout if its utilities could not demonstrate the financial security to procure power from outside sources.

The risk of complete, long-term power outages has been underplayed in the press and in discussions of the restructuring problem, perhaps because the problem was averted. Yet what would happen if power suppliers suddenly judged a utility that serves 10 million people as too risky to sell to? Clearly if enough power was withheld from the market, the shortage that resulted would cause a blackout. Not just a several-hour blackout, but a complete, extended systemwide blackout.

What would it take to turn the power back on again? Someone would have to assure power providers, not one-by-one but as an entire market, that a reliable buyer was available. This could have been very difficult to do without essentially a government takeover of the electricity system.

Here we see another weakness with economic theory: in theory, an individual or a corporation that desires to pay for something can always do so, even if it has to borrow money. A utility that has a legal obligation to serve its customers and the ability to collect its costs through sales would always, in theory, be able to procure power. But in the real world, there are limits to both personal and corporate credit, even if those limits seem illogical.

In this case, the limits were completely nonsensical; it was inevitable that the PUC eventually would allow utilities to raise rates and pay all obligations for power purchases if the only alternative was to let the entire electric system collapse. So the revenues of power suppliers were assured. Yet the real markets didn't see it this way.

The State Steps In

Perhaps recognizing this immense risk, at this point the state stepped in. California's action was probably critical in terms of preventing even worse damage than occurred, but in terms of the stabilization of power markets it was too little, too late. In January 2001 Governor Davis signed Senate Bill 7X that enabled the Department of Water Resources (DWR) to procure power instead of the utilities, who had lost credit-worthiness as buyers. The DWR negotiated high-priced, long-term power contracts that it could actually get (unlike the investor-owned utilities) based on its (at that time) high credit rating. Perhaps these high prices were inevitable given the experience of the past year.

While the state acted assertively to provide the financial security needed to keep the lights on, there were obviously some government officials who felt that the utilities, not the consumers, should continue to pay the high prices, even if they were to go bankrupt doing so.

So two-and-a-half months later, in April 2001, despite this state intervention, PG&E declared bankruptcy.

The True Causes of the California Energy Crisis

In summary, the key causes of the 2000 California energy crisis were:

- Investments in efficiency, renewables, and conventional power were slowed and disrupted by uncertainty created over restructuring and the lack of incentives for anyone to pay for electric reliability.

- Distrust of utilities both by the right wing—which worried that utilities were big monopolists that thwarted free markets—and the left wing—which worried that utilities were greedy capitalists that try to raise rates and enhance profits excessively—combined to take utilities out of the picture as a possible source of the solution. Utilities did not have the regulatory or financial support to build power plants, contract for long-term power, or even maintain (much less expand) efficiency programs to obviate the need for power.

- Untested economic theories were implemented without consideration of the weaknesses and limits of pure market forces in a real world of business and politics.

- Finally, the financial part of the crisis was exacerbated by the aversion on the part of practically all interest groups to pay high electric rates or to allow utilities to charge them. Again, different stakeholders had different reasons, but no single stakeholder wanted to be in the political position of advocating that businesses and consumers pay more. This should not be a surprise. Yet it means that market forces are constrained by political forces, a fact that flies in the face of economic fundamentalism. Markets depend on the rule of law. In a democratic society, the rule of law will not allow situations to persist that are perceived as unfair.

This story did not stop with the signed long-term power contracts. While California had lost $15 billion due to high electricity prices in 2000, experts predicted without serious challenge that the excess cost for electricity would be $40 billion(!) for 2001, coupled with forty days of rolling blackouts.

Once the seriousness of the problem in the summer of 2000 became apparent, state government worked in cooperation with utilities, non-profits, the legislature, and the private sector to design an emergency conservation and efficiency program that could keep the lights on in 2001. This program combined information and "jaw-boning" by the state to individuals and business owners to conserve energy, a four-fold expansion of funding for utility energy-efficiency programs (paid for by the state general fund), an emergency increase in building and equipment efficiency standards, and an accelerated licensing procedure for new power plants.

The efficiency programs were remarkably successful: peak power demand in California was reduced year on year by 12 percent, 9 percent, and 8 percent in the three peak months of 2001.[5] This reduction came despite somewhat warmer weather and continued economic growth. And the largest percentage decrease occurred before electric rates increased—in stark contrasts to the fundamentalist belief that reductions in consumption are caused primarily or even exclusively by price increases. Such a reduction is unprecedented. The only years in which electricity consumption fell more than 5 percent nationwide in the past century were the worst years of economic depressions, 1921 and 1932; and even in the worst year, 1932, the decline was less than 10 percent.

The success was nearly total: no rolling blackouts occurred after 2001; and of the additional $40 billion of cost predicted, none materialized.

And much of the savings persisted. As of August 2005, the highest level of peak power for the state has yet to surpass the peak of summer 1999, even though California has since experienced 4.5 percent annual economic growth.

The Surprise Impact on Economic Fundamentalism

The claims of the proponents of deregulation turned out to be illusory. The promised reductions in price hardly occurred during the years when the program was successful, and were wiped out more than one-hundred-fold by the first year of the crisis. Even the main proponents of restructuring, the large industrial customers, lost as a result of the headlong rush to deregulate. Following this experiment, the state is left with a legacy of financially weakened utilities and high-cost, long-term electricity contracts. Electric rates increased some 40 percent in 2001 and are expected to stay at that level indefinitely. Also the rate increases were not confined to California: Nevada, Washington, and Oregon also suffered 33 to 40 percent rate increases as a consequence of the California crisis.

Worst yet for the free-market fundamentalists' credibility, the implementation of competition resulted in all sorts of irregular behavior by power suppliers.

And interestingly, even the power suppliers didn't benefit from the crisis in the long term because they suffered a bust following the temporary boom, and, in the cases of Enron and Mirant, wound up in bankruptcy courts. All of the other energy companies were also financially weakened, which leads to the possibility of an "echo" crisis as demands for power plants continue but no one has the financial strength to actually build and finance them.

Most importantly, the strong assurances by proponents of the restructuring system that they had studied the successes and failures in other countries, and had gotten it right, were totally blown out of the water. We had an immense experiment in how much to trust the judgment of the free-market fundamentalists and discovered that most all of their claims exploded in our faces.

Yet oddly these claims did not explode in the faces of their proponents. Surprisingly, these ideas continue to be trotted out by some of the same interests, who seem to have entirely evaded the responsibility for their radical experiment to establish new markets for commodities that may not be market-compatible, and to impose specific designs and rules of the road for markets—which turned out to fail.

Again, some have argued that the failure of the economic fundamentalists to get everything that they wanted in terms of the structure of the program (in other words, they did not get the ability to pass on price increases to retail) shows that we needed the courage to deregulate more fully rather than partially.

However, this would not have changed things much, as a further discussion in Appendix A reveals. The excess costs would have been slightly lower, yet still punishing to the California economy. In other words, the experiment with market forces, which was undertaken for the sole purpose of lowering costs and making the state better off, would still have resulted in higher costs and would still have left everyone worse off.

Yet in fact, the proponents of restructuring got almost everything they wanted. To the extent they fell short, no one ever succeeds in a democratic political structure in a system redesign as big and as complex as electricity and has every possible detail go precisely as desired. If a system exhibits such all-or-nothing characteristics, it is simply inconsistent with the democratic legal structure that underpins a free-market system.

Thus far I have discussed the dangers of economic fundamentalism as an advocacy strategy for those opposed to environmental regulation. However, the California restructuring story is different—it illustrates what can happen when economic fundamentalism is used to start something rather than to stop something. It shows just how large the damage can be when flawed theory is used as the basis of real market design, and it demonstrates with actual experience what some of the flaws are and how they led to the observed failures.

California's story shows that a "free-market reform" designed to cut prices through enhanced competition and to increase growth in fact had opposite result: higher-cost electricity, business failures, and reduced growth.

The rejection of market fundamentalism is different than the rejection of markets. It means that economic theories must be checked against real-world observations. If ideal markets lead to optimal solutions, but real markets fail to follow this ideal model, we can start to look at how they depart from optimality. This kind of observation can lead policymakers to exploit the strengths of markets, overcome some of their weaknesses, and use alternate approaches to correct for the cases where markets fail more fundamentally to produce a good outcome. (This discussion is explored next in Chapter 6.)

6 | How Markets Fail

Despite my strong exception to the philosophy that markets always achieve economic development goals, I believe it is nevertheless true that markets tend to work better than any other mechanism that has been advanced to make economy-wide decisions. Markets allow bottom-up, or distributed, decision-making, which is less prone to error and more conducive to innovation than top-down, or centralized, decision-making. Markets provide for competition, where better ideas or processes can drive out worse ones and new, more productive approaches can replace obsolete ones. Therefore one should look first at what a market-based solution would be and then make strategic adjustments to achieve the goal of economic development when markets fail to work in practice as well as they work in theory.

So it is important to understand how and where markets fail in order to see what the consequences of these failures are. An understanding of the failures provides policymakers the basis on which to decide how to overcome them—or, if that's impossible, how to compensate for them.

What Prevents Expected Results

To design markets whose ground rules are more effective, or to correct current markets wherein failures have occurred, has been the goal of much of the policy work on energy efficiency. Analysts have looked at what they call market barriers or market failures—practical problems that prevent markets from delivering the expected optimal results.

Unfortunately economic fundamentalism, to a greater or lesser degree, colors many of these analyses. This flaw shows up in policy analyses that suggest that unless we can understand and document exactly how and why markets fail, we have to assume that they work. Such a formulation omits the possibility, which is often seen in practice, that markets fail to provide the optimal solution although we can't explain why.

Such an omission is a serious problem because often we can solve a problem in environmental policy even if we don't understand or can't explain why markets work. One example of such a solution—rebates for purchase of energy-efficient products—is presented below.

The Effect of Rebates on Market Decisions

In the 1990s we observed that cost-effective technologies for refrigerator efficiency were not offered in the market. Consumers were unable to buy the products because they were unable to find them, and retailers failed to stock them because there was insufficient consumer demand. The problem extended also to the manufacturers: producers wouldn't make the efficient products without orders from their retailers, but retailers could not order something that wasn't produced.

Some utilities found that rebates of about $50 solved the problem; and utilities that offered rebates got almost half of their markets to buy the efficient options.

However, evaluation studies revealed an interesting result. Consumers almost universally said that they understood the value of efficient products and would have bought the more efficient products even without the rebates. Retailers said that they just responded to market demand—the rebates made no difference. And manufacturers said they produced products that were compliant with the specification for the rebate anyway.

Only one problem obtrudes into what is otherwise a consistent picture. In areas without the rebates, virtually none of the efficient products was sold. And during seasons when the rebate was withdrawn, sales of the efficient products plummeted. So we could recognize failed markets and we could see how to fix them. Yet it would be impossible to say, based on evidence, how markets had failed.

So an excessive focus on the identification of specific mechanisms of failure in markets causes policy makers to miss opportunities to correct them, and injects an element of economic fundamentalism into the discussion.

New Terms to Define How Markets Fail

To avoid this problem and clarify future discussions, I use different terms to distinguish how markets fail. Market barriers, which I define as relatively simple problems that can be identified and corrected; market failures, which are more complex and may not be easily correctable; human failures, which are more fundamental and may be incapable of correction in a way that follows the economic models; and institutional failures, wherein the structure of economic decision-making departs from market-compatible relationships. Each of the four basic reasons why markets fail is examined below.

Market Barriers

Whatever their weakness as instruments to provide optimal environmental and economic decision-making, markets are still the fundamental basis of most economies in the world. In general, markets clearly work better than virtually any alternative economic decision-making structure. So it behooves us to scrutinize how markets fail because to do so provides guidance as to how these failures can be corrected.

In some cases, markets would work save for the existence of what I call market barriers. The concept of a market barrier is that if policies can be developed that overcome the barrier, markets will automatically shift toward the right decision.

An extensive amount of literature on market barriers, particularly for energy efficiency, is available, the scope of which goes beyond this book. For our purposes we can use examples of some of the more important market failures to identify how to avoid or overcome them.

The Barrier of Limited Information

One of the critical market barriers is lack of information. In the case of energy use, a consumer inherently has no idea what kinds of energy costs a given decision will impose on her unless a way to rate energy use is identified. For automobiles, this barrier was overcome, at least to some extent, with the establishment of the miles-per-gallon sticker. This involves the creation of a test protocol and the regulatory requirement that manufacturers test all products in accordance with that protocol and publish the results.

A similar approach was established with the yellow Energy Guide label in the United States that rated appliances.

A different approach to labels in the European Union (and apparently much more effective one) uses a rating system based on letter grades A, B, C, D, E, and so forth. Another successful approach found in Australia and New Zealand uses a star system that awards an increasing number of stars on the labels of more efficient products. For the most part, however, people are unaware of what products or activities consume lots of energy, and which ones are scarcely worth the worry. And even if information is available, such as on the Energy Guide appliance stickers, consumers may be uncertain about the stickers' purpose, or more fundamentally, they may fail to see the significance of the information.

In general, one way to overcome lack of information is to provide nationally or globally consistent rating systems and require accessible information; another is simply to provide educational materials on the subject of energy efficiency (or any other environmental subject). Many utilities and nonprofit organizations, as well as government agencies, take this approach.

Unfortunately it is difficult to measure scientifically the

effectiveness of these information programs, so at times it is challenging to justify politically incurring their cost. So while it may be possible to overcome this market barrier, the value of doing so is not evident.

The Barrier of Split Incentives

A second key market barrier is split incentives. Consider, for example, an office building in which a different tenant leases each floor. The landlord is responsible to upgrade the building and make it more energy efficient (or not), and it is the landlord who incurs all energy upgrade costs. However, the tenants typically pay the bills, which are divided among them according to how many square feet each tenant rents.

The landlord receives the same net rental rate each month irrespective of utility costs. So if the landlord invests in energy efficiency, the tenants themselves will share the collective benefits, while the landlord will receive zero return on the investment.

One might think this easy to fix. Yet suppose a tenant makes the efficiency investment for the leased floor only. A split incentive is still possible. The electric bill is for the entire building, so the tenant who invests in efficiency will have to share the savings with all other tenants. Worse yet, the efficiency investment might have a twelve-year life, but the tenant has only three years left on the lease.

Split incentives can occur between individuals, among corporations, or even across management divisions within the same corporation. For example, many or most corporations have a capital budget that must pay for energy-efficiency upgrades, as well as other sorts of capital upgrades, and an operations budget that pays energy operating expenses, including the cost of such energy-related activities as light-bulb changes.

Such budgets mentioned above are not interchangeable; they may even be under the management of different individuals or departments. Despite the physical opportunity for an energy upgrade to return all of its initial cost in two years—or even the first year—the capital budget manager still would be uninterested in undertaking it. The project's costs would be a drain on the capital manager's budget and the benefits would all accrue to somebody else in the organization.

Economic theory states that split incentives should be unable to persist. Consider the landlord/tenant problem above. If one landlord has an energy efficient building that costs $50 a month for utilities while a neighboring building's cost is $150 month, the first landlord ought to be able to charge $100 a month more rent than the second. So in principle this barrier is easily overcome. Yet in practice this almost never happens.

The Barrier of Uncertain Performance

For environmental improvements that involve new technologies, which is typical of energy efficiency as well as other areas, performance uncertainty is an additional market barrier. New products may or may not perform as expected; they may have side effects on other performance attributes about which the buyer, who may never have tried one before, would be nervous. Performance-uncertainty barriers can be overcome if a trusted source can vouch for the acceptability of a product, or provide reliable guarantees that ensure against the worst possible outcome of using the new technology.

More Serious Market Shortcomings

Some policymakers tend to put everything into the category of market barriers, which suggests that government interventions in the marketplace should be limited to those that can identify specific market barriers and then figure out ways to overcome them. Economic fundamentalists might even suggest that once this had been done (if they even accept the reasonableness of it at all), markets will then begin to function perfectly and the intervention can cease.

However in other cases, policies have demonstrated their effectiveness to produce the right economic answer without any explanation as to how or why. Sometimes the effort to identify market barriers and correct them amounts to little more than a paper chase because we already know how to solve the problem.

A more serious critique of the concept of market barriers is that sometimes the problems are much more fundamental than mentioned thus far. The previous discussion has focused on areas where markets fail to perform perfectly, but where relatively simple fixes could correct the problem. More fundamental failures of the market to work without strong public policy intervention, or, in some cases, its failure to work correctly in any event are discussed in the next section.

Market Failures

A more fundamental problem with markets is what I refer to as market failures. Such a problem refers to areas of the market that fail to provide the answer that maximizes welfare either for reasons we have yet to understand or for reasons that are so fundamental to the structure of markets that they are difficult to overcome. The existence of market failures, particularly those that cannot be corrected, implies the need for alternative policies that can give the result that a functioning market would have given, or can cause policy-directed outcomes that are politically determined to be good for everyone even if they're independent of a market process.

The Failure of Diffuse Decision-Making

An important market failure in the commercial building sector is diffuse decision-making. For markets to work properly, a single individual or organization must be held responsible to make decisions, such as those that concern energy efficiency. But the process to develop a commercial building is generally so complex that no one individual or corporation can make a firm decision. In the absence of such authority, the safe thing is to agree to conventional practice and dismiss anything that is unconventional.

For a commercial building, energy-efficiency innovations must be approved by the architect, in most cases the engineer, possibly also a lighting designer; they must be acceptable to the contractors within each of a number of trades, of course the developer of the building—the corporation that owns it as it is being built—has primary decision-making authority, but the developer will avoid doing something if it is not acceptable to the two prospective eventual owners (or even if the developer intends to own the building and operate it, the company will still have to consider its market appeal later on), the construction lender, and the mortgage lender. Changes from normal practice will also have to be acceptable to leasing agents and prospective tenants, and will be noted and considered by appraisers.

With all these different decision-makers, a change from conventional practice is exceedingly difficult: it is virtually unheard of that all persons come together in order to agree to something. One guaranteed meeting place, at least by representation, where they are "in the same room, at the same time" is at the building codes official's desk where the designs of the building are approved. This is one of the reasons that energy-efficiency codes have been so effective: the code provides a virtual room where all decision-makers have to agree on specifications.

The Failure of Private-Sector Regulations

Another market failure is based on the power that players in the market often have to define the rules by which the market works. Chapter 4 discusses private-sector regulation and how it tends to favor economic incumbency. I elaborate on this issue below. Yet because of the sheer number of regulations needed in a technology-based economy and the concentration of technical knowledge among the producers of the technology, a permanent danger of conflict of interest in the way markets work presents itself. It is difficult to imagine how to overcome this problem, so it is likely to continue as a market failure.

The problem of private-sector regulations that constrain markets is not limited to formal regulations. It can include informal rules and simple business practices that suppress free decision-making in the market.

The first example we have seen is the informal regulations that the financial services industry places on credit. The final stage in the California electricity crisis was precipitated when the biggest investor-owned utilities did not have the financial strength to buy power. The state had to step in and use its high credit rating for Californians to procure power at all. The business reality of credit-worthiness is contrary to economic theory. In theory, if a business, say a utility, wants to make a transaction and is willing to pay, it can, particularly for a commodity like electricity. But in the real world, standards of practice in the financial industry prevent these transactions. The government does not require these standards; business voluntarily accepts them. However, they stand in the way of how the ideal free market functions.

Similar problems also impede business investments in energy efficiency. In theory, a corporation with an investment opportunity at 30-percent return and a borrowing cost of 5 percent would always be able to make the investment. But in practice, rigid limits to debt relative to income in the corporate finance field exist, and often the manager who is faced with the 30 percent opportunity can't exercise it because it puts the company over a limit and hurts its credit rating. In many cases, these corporate borrowing limits result in a system that allocates capital budgets to each department. (Allocation or rationing is not a market-compatible system; a market-based system would send capital to the places where the return is highest, based on competition.) In this case, a department manager who discovers an energy-efficiency opportunity could not take advantage of it no matter how high the rate of return!

A similar policy limits home ownership. If a family with an annual $35,000 income wants to buy a $300,000 home, it will find it impossible. No bank will loan them the money. While there is no government regulation to that effect, the lending industry is as tied to its internal regulations as it would be if the policy were law. The reason for this industry practice is that the cost to repay the loan and taxes in this example amounts to 60 percent of the family's income, and statistically most families can't afford to pay that much.

Yet suppose your family is somehow able to successfully meet the financial challenge. Under the informal private-sector regulation of almost all lenders, the answer will still be "no"—even if you can demonstrate that you currently save enough money every month to meet the payments. (Suppose, for example, that your family had previously been earning $15,000 and keeping out of debt. In this case, you would be able to spend the entire extra $20,000 all on the new house.) These informal regulations deny consumers the access to desired opportunities—even when they can afford them. And just because the regulations are informal they are no less restrictive. Consumer sovereignty and free choice in the real economy are limited compared to the theoretical economy. These restrictions are an example of market failure.

Informal regulation is a major barrier to the use of smart growth: neighborhood developments that rely on compact development plans and the use of mass transit to increase location efficiency and reduce transportation costs for residents. For the developer, a smart growth subdivision in the suburbs or an urban infill project both fail to conform to the standard requirements for a financeable project. Instead of the desired automatic loan approval for the project, the developer has to instead make an individual appeal to the lender, and often is faced with a higher interest rate or even outright rejection.

The prospective homeowner faces a similar problem. The bank that will write a mortgage for $1,300 a month for a family that applies to buy a home in a sprawl subdivision where its cars will cost $700 a month (for a total of $2,000 in monthly obligations for transportation plus housing) will refuse to finance the family's alternative choice of a house in a smart growth neighborhood with a $1,600 monthly mortgage and transportation costs of only $300 a month (for a $1,900 monthly total). The Location Efficient Mortgage®—a system that credits a prospective borrower with the value of their transportation savings, $400 a month in this example, when qualifying them for a loan—was created to overcome this specific market failure, but the lending industry has accepted it only in the most limited way.

Informal regulations affect real estate development more broadly. Because large corporations that do not tend to keep the same property for the long term own much of the property in the United States, the industry has found it convenient to develop new areas according to standardized formulas. If a property falls into one of the roughly eleven categories of building and if it has typical characteristics for that category, then it is easier to sell, and the value will be higher. If it fails to meet this informal regulation, then it will be more challenging to sell.

Both formal and informal private-sector regulations can cause market failures. Alternatively, the lack of regulation may be a market failure. For example, many homeowners are uncomfortable when they select a contractor to work on their home, as they fear incompetent work or unfair pricing. This concern could be alleviated by a ranking system, either private or governmental, that evaluated contractor performance on a quality scale. Yet such rankings are not offered.

Some of the failures of informal regulation are due to its actual informality. The electrical blackout that hit some 50 million people in the northeast United States in the summer of 2003 resulted due to a failed enforcement of an informal regulation. A discussion of how this happened follows.

For years, utilities in the United States were vertically integrated monopolies. They displayed much social cohesiveness in their actions, perhaps because their ability to compete was so circumscribed.

Utilities have always bought and sold power over long-distance transmission lines. This reduced the cost to assure reliability. Yet systems linked by power lines are also linked in their reliance on each other. If power deficiencies exist on one end, both systems are at risk for blackouts.

The industry has recognized such interdependence, and a long history of voluntary agreements has emerged on how to share power. These informal regulations had no formal enforcement mechanism. (In fact, for decades one wasn't required.)

However, in August 2003, such voluntary agreements broke down. Some suppliers acted to maximize their individual profitability, rather than follow agreements that called for suppliers to share the pain for mutual benefit, and did not provide sufficient "reactive power" (a kind of power for which producers are unable to charge). The result was a region-wide blackout that cost the United States and Canada billions of dollars. In response to this failure, Congress included mandatory regulations on how utilities and power generators must cooperate as part of the Energy Policy Act of 2005. These mandatory regulations superseded the informal regulations that had failed to work. The regulations were adopted without controversy.

The Failure of Price Competition for New Products

Another sort of market failure is based on the fact that the technological basis of most of what the economy produces yields increasing returns to scale. Mass production lowers prices. This effect is compounded for new technologies with the learning curve effect: continued production of the same type of product leads to further price decreases due to increasing experience and learning by doing.

So a new product—for example, a more efficient air conditioner—will cost more not only because it uses more advanced technology but also fewer of them are sold. A consumer who individually seeks energy efficiency will find the options are more expensive than if more people chose it. This dynamic creates a vicious circle: a product that would be cost effective if everyone wanted it instead is excessively expensive, which means few people will buy it, and the result is an even more expensive product.

Equipment manufacturers have pointed out this market failure. They told utilities and energy agencies that if they want to promote the more efficient products, they need a single target level to shoot for that is uniform throughout the country, and that is fixed for several years. They also voiced a need for financial incentives to sell the product.

Economic theory states that for products with a learning curve, the correct price is that of the last unit of production when the industry is far along the learning curve. Yet this only works if each company can hold on to the profits from the learning curve. If information leaks from one innovating company to its competitors, then the producer will need to charge higher prices in order to protect its profits. The new product therefore will be overpriced. This appears to be the case for environmental technologies.

The examples above illustrate how government interventions of the type used in energy-efficiency policy, and more broadly in environmental quality policy, overcome fundamental failures in the marketplace that prevent the determination of an economically sensible answer without these interventions. In some cases, the examples highlight that government regulation, rather than standing in opposition to how a competitive free-market economy functions, may actually enhance that operation. To the extent this is true, it is clear how environmental regulation policy in general and even environmental regulation can promote economic development.

In the last case, the market failure justifies government involvement in the financial support for emerging technologies that will be cost effective when they are far enough along the learning curve.

Human Failures

Markets consist of human beings who act in social and psychological, as well as economic, contexts. Humans do not always follow the principles of welfare maximization, even for themselves, much less for the organizations they represent.

If the psychology of humans predisposes people to act in ways that discourage innovation, then government policies that actively encourage it are more likely to be needed, and far more likely to be an effective agent to enhance economic growth.

The types of human failures that would contribute most to stagnation in the economy and to the need for outside regulatory or policy pressure to induce the economically expected optimal result are factors that reinforce the status quo.

Managers will resist innovation if they are in a culture or peer group that does not support it. They will resist innovation if they associate it with unfamiliar people or positions. Managers will also resist innovation if they are risk averse or care more about how to preserve current economic benefits for their firm rather than how to maximize future profits.

Unfortunately, all of these phenomena seem to be happening concurrently.

Human Failure Under Peer Pressure

Peer pressure is a form of human failure not often discussed, in part because it is difficult to measure.

Managers in a particular industry tend to meet with and socialize with members of their peer group in the same or competing industries. Many industries sponsor regular conferences or information-sharing meetings that attract attendance that crosses company lines but stays within a given profession.

So even when companies compete with each other, this competition is tempered by a group identity (as members of a particular industry) or by bonds of friendship or collegiality that transcend strict competition.

The result, from my observation, is that in many cases companies will compete vigorously in some areas, such as price or consumer-responsive features, but will refrain from action in other areas that their peers would view as too disruptive.

Such behavior tends to suppress innovation.

Peer pressure represents a psychological force that encourages people to make decisions that depart from what they would otherwise know to be rational, were they to think about it dispassionately.

It's incorrect to say that peer pressure is always a bad thing.

Peer pressure can lead to informal regulations, or equivalent self-reinforcing expectations of honest behavior, that are necessary to make markets work. Peer pressure can impose social responsibility on corporate decision-making.

Yet peer pressure does, more often than not, hold back innovation and creativity.

The Human Failure of Not Paying Attention

Another human failure is what is referred to in the market barriers literature as bounded rationality. Bounded rationality refers to the decision-maker's inability to focus attention on all aspects of profit and loss, but rather concentrate on a few critical areas only. Because energy cost is seldom the biggest or even second-biggest cost a manager is responsible for, it may be ignored completely.

Bounded rationality is a difficult problem to correct in markets. People generally tend to focus on the biggest issues and ignore the smaller ones. I previously noted that few CEOs read fifty-page decision memos, and simple market-based programs are unlikely to change this.

The human failure of not paying attention is a particularly severe problem for the environment. No market for environmental quality exists: environmental aspects of products or of the production process are secondary attributes of other products or processes. Even energy efficiency, which has a direct relationship to costs, is always merely one attribute of a multi-attribute decision. And it is rarely the most important or even second-most-important attribute. If people have limited attention spans, environmental features will always fall off the table—ignored as busy decisions-makers focus their attention on top priorities.

Other human failures cause people to choose options that are clearly contrary to a rational model, and contrary in a way that consistently discourages risk taking and innovation. Three of these characteristics have been characterized precisely and analyzed: loss aversion, risk aversion, and status quo bias.[1]

The Human Failure of Loss Aversion

Loss aversion was first noted in scientific literature in 1980 in a paper that demonstrated that people typically are much more averse to giving up a possession that they already have than they would be willing to pay to purchase it in the first place.[2] Numerous experiments that present people with hypothetical choices confirm this effect.

So if the loss-averse individual who manages the production line for an energy-reliant device is offered the option to pay additional money for a redesign of the assembly process, coupled with the likelihood that the resultant process change would make more money, the concern over the potential loss is rated more highly than the excitement over the potential benefit.

If the cost of the improvement is not known, the aversion to loss will obviously increase. And given the structure of industry, in which process engineering is almost never done and in which there is no experience to validate the idea that even a successful product redesign that saves energy at high return on investment (ROI) to the customer will lead to increased sales or profits for the manufacturer, loss aversion will be the dominant human response. In fact, the market failures described above will lead the manager to assume the reverse: that even a successful manufacturing change will not likely lead to increased sales or profits. A manager's first impression will be to resist change and to assume that it will affect the company adversely.

In other words, the other failures of the market serve to validate and reinforce the natural tendency to loss aversion.

Of course, this concern is magnified if the manager would be individually responsible for the loss but would have to share the gain with the rest of the company. Yet even if the manager is compensated in a way that she shares equally in the benefit to the corporation, this type of human irrationality will cause her to under-invest in innovation compared to making a rational tradeoff of risk and reward.

Note that such a problem affects more than profit-making

corporations. Managers at nonprofit organizations and government officials will make the same mistakes.

The Human Failure of Risk Aversion

A similar but distinctive phenomenon is risk aversion. The example, given by Professor Daniel Kahneman of Princeton University, who shared the Nobel prize for economics in 2002 for his work on this subject, presented the following case: "Would you accept a 50-percent chance to win $150 along with a 50-percent chance to lose $100?"[3]

The likely outcome of this coin toss is a gain of $25 each time it is taken. Yet most people would not accept the offer. Interestingly, the answer is not dependent on a person's wealth.

Similar experiments show that if presented with an option to accept a painful or expensive inoculation that offers 100-percent protection for a disease that's only 1-percent likely, far more people will do this than when offered a similarly painful or expensive inoculation that provides 50-percent protection against a disease that has a 2-percent chance of occurrence.

Rational consideration shows that the two cases are absolutely identical. (To see this, imagine that the 2-percent risk consists of two separate diseases, each of which has a 1-percent risk, and the inoculation is completely effective against one but totally ineffective against the other.)

Numerous other experiments by several authors show the same phenomenon. In general, people prefer to avoid loss more than to secure gains by a factor of 2 or 2.5. Furthermore, people's responses are not proportional: a potential gain twice as large is less than twice as valuable as the base case.

The Human Factor of Status Quo Bias

A third psychological factor is called status quo bias, which manifests itself when a decision-maker is presented with complex choices: she or he tends to prefer an existing situation. The more complex the choices are, the stronger the status quo bias.

An example from the research is an experiment in which the subjects are told that they have inherited an unexpected large sum of money. Each receives, say, four choices of investments for their windfall. Another group receives the same question, but with the additional information that the deceased had previously allocated his money in a specified way among those four choices (before the estate's executor cashed out the investments). The second group was far more likely to duplicate or come close to the original choices than the first group. And the greater the number of choices, the greater the tendency to leave the situation as it was.

Taken together, these three biases discourage managers from investing in process improvements unless the benefits are perhaps eight-to-ten times the cost. (Only one of these three irrationalities would cause an under-investment ratio of 2 or 2.5; if the others have similar distorting effects, we would get a range of about 8 to 10.) Particularly as the proposed change departs further from the status quo, and is increasingly complicated, managers will adhere to standard behavior rather than innovate, based on the way the human brain functions.

In this case, government interventions that encourage more innovation are likely to produce substantial economic benefit.

These human failures may not be so important when we consider the actions individuals take on their own behalf. If someone is risk-averse in his own behavior, this behavior could be described by an economist as that of a rational person who has a set of preferences that differs from what one might expect.

In other words, if an individual acts on his own irrational preferences, a Pareto-optimal result may be achieved if the consumer gets what he irrationally thinks he wants rather than what he truly prefers.

Yet when the same person acts in a risk-averse fashion while an agent for a corporation, the result is different. The corporate shareholders don't want irrational behavior, and such behavior certainly is not consistent with how to maximize profit. Human failures that occur when the person acts as an agent for others are serious barriers to a properly functioning market.

Institutional Failures: Trade Associations and the Politics of Environmental Protection

The role of trade associations in the real economy of the United States embodies another sort of market failure: the potential for undue influence of the larger economic actors—the economic incumbents—in setting forth the rules of the marketplace.

Chapter 4 illustrates that markets can only function when the government or another institution establishes ground rules that define the process of market competition. We have also seen the importance of regulations that define the basis on which markets allow competition for goods and services.

Yet what if the economic units, such as corporations, not only follow the rules, but also actually write the rules? Couldn't this be a source of resistance to competition?

The Role of Trade Associations

Trade associations, particularly on a national level, play a vastly underappreciated role in the establishment of the rules under which the markets for their products operate. Could this role undercut competition? Strong theoretical reasons exist to expect this result, although almost no study is available to confirm or contradict it.

In my experience, the role of trade associations has included advocacy and regulatory decision-making to protect economic incumbency and reduce competitive pressures on their members. My experience does not suggest that trade associations do anything illegally anti-competitive, but it does show that the current system does consistently diminish the role of market forces, and therefore is a failure of markets.

I had my first experience with trade associations in 1986 during debates in front of regulatory agencies over the issues of whether to have building energy-efficiency standards and appliance efficiency standards, and if so, over what level of efficiency to choose. The roles and positions of the participants were remarkably simple at that point: the regulatory agency generally proposed fairly mild regulations; environmentalists advocated that they should be strengthened; and the industry side argued against the need for all regulations.

Industry spokespersons included both representatives of individual companies and of their trade associations. The trade association had a natural and seemingly unremarkable role because, particularly at that time, industry had a common position on the issue, and so you would expect an association that represents that industry to be able to present the group position in a cogent way.

However, the role of trade associations remained dominant even when the companies in the industry did not all agree. When environmental advocates and industry declared a truce in their long-running battle over appliance efficiency standards in 1986 and began to negotiate with each other, the negotiations were structured in a bilateral form. The three trade associations that represented appliance manufacturers (one association represented one set of products, another represented a different set) sat on one side of the table and the Natural Resources Defense Council (NRDC), which was representing both environmentalists and advocates of strong standards, sat on the other side of the table.

The structure of the negotiations—an industry side and an efficiency advocate side—may seem natural at first, but it should also raise serious concerns for believers in a competitive marketplace. The structure was such that all of the companies within a particular industry had to agree to a common position before an offer or a response to an offer from the other side at the negotiating table could be presented.

If one company, for example, had an interest in a stronger standard and another company had an interest in a weaker standard, the former company's position was subordinated (with its active approval) to the group position. Notwithstanding that at least one company actually supported stronger standards, the peer pressure, or loyalty, or group identity, of industry led to their negotiating as a bloc, rather than as a representation of individual interests.

The trade association–enforced group cohesion almost led the 1986 negotiations to fail because the trade association apparently couldn't agree to a proposal by the efficiency advocates unless all of their members agreed. (I am assuming that they followed this rule based on what I observed, but obviously I was not informed of their internal processes.)

This process leads to a least-common-denominator approach when the issue concerns the use or development of new technology. In a multilateral negotiation, the final consensus might have been around a more technically demanding standard that would benefit the companies that were technology leaders at the expense of those that lagged. One would expect that the leaders would be happy to make alliances with parties outside their industry to promote standards that would improve their market share as well as increase their profit margins, but in almost all cases, beginning with the 1986 negotiations, companies preferred to work within the trade association context and forgo competitive advantages that could have been exploited had they negotiated or lobbied as corporations.

So just how typical is this behavior?

Unfortunately, discussion of trade associations and their role in policy setting is almost entirely lacking in the academic literature. Indeed, even to collect the basic information on them appears to be a daunting task. Despite the obvious facts that one sees when walking the streets of Washington, D.C.—that trade associations are a major presence in the city—no information seems to exist on trade associations. No one seems to know the total number of them, how many employees work for them, or to what extent they lobby. Considerable additional study is seems necessary.

Such a lack of information is of concern because my experience has been that trade associations play a major role in economic policy in the United States, not only through their role as writers of private-sector regulations, but also through their role as shadow legislators. Trade associations lobby Congress to pass legislation that defines the rules of markets and the extent of competition. They also lobby administrative agencies on the content of regulations that affect their markets.

Trade Associations' Influence on Law Makers

Another poorly noted and largely undocumented phenomenon is the additional role that trade associations have taken in legislative deliberations: as power brokers and dealmakers. Trade associations often negotiate deals in which one association supports an interest irrelevant to them, but important to another association, in exchange for that association's support on a different issue.

Such reciprocity is important more for its economic significance than for its political relevance, even though the political importance should not be underestimated. When trade associations make deals on national economic policies and national environmental policies with economic effects, this process undercuts competition and market forces.

A deal by a trade association that acts individually would only be undertaken rationally if it advances the industry in its entirety. To advance the entire industry is something that is challenging to do in a way that enhances competition, but easy to do in a way that suppresses it. (Think of the issue this way: if a policy enhances competition within an industry, it will almost always produce some winners and some losers. The losers would usually stop their trade association from promoting that policy. So the only policies a trade association would feel comfortable in advocating are those with all winners and no losers, or those in which the losers accept their losses in the interests of peer solidarity. Neither category suggests an opening for policies that are pro-competition.)

When trade associations act as dealmakers, it is almost inconceivable to think of how these deals could do anything but suppress and evade competitive forces in the economy.

Trade associations, while steering very clear of any legal anti-trust violations, can nonetheless engage in tacit collusion to protect the interests of economic incumbents against the risks of competition from new entrants.

In addition to the political activities discussed above, trade associations can provide unified industry input to other organizations' private-sector regulations or even governmental regulations. In any of these cases, an obvious desire would be to structure the regulations to make it easier to do what the bigger and more established companies currently do and to make it more difficult (whether by plan or inadvertently) to do things differently. This in itself tends to favor the status quo over innovation and established companies over new entrants.

The Effect on Markets

The policies and practices mentioned above have the tendency to diminish the role of real markets in the economy and to thwart competitive forces. Therefore policies designed to overcome these political dynamics will, at least incrementally, free American markets further and promote competition, as well as promote innovation.

Actually, it is possible to reach similar conclusions from simple game theory. The game starts with many companies within an industry that compete in a marketplace wherein at-issue government policies determine some of the rules. Clearly, in either case, it is to a particular company's benefit if it can access information about the policy debate in Washington, and how it might participate to protect or advance its interests.

Yet information is difficult and expensive to obtain in Washington. Most companies, that, as we have noted, are not managed in a way that offers large strategic planning resources to executive management, would first find it advantageous to pool resources and provide information through a trade association with a professional staff in Washington.

So a new trade association comes into existence and hires a staff. The staff initially listens more than it speaks, and passes information back to its members. But the staff quickly discovers that if the trade association members all have the same position, it is able to offer a value-added service to membership by advocating that position on behalf of the whole industry.

A few successful advocacy experiences with common industry positions makes the staff realize that it has more power—that its own professional careers are advanced—if it has a common industry position from which it can advocate. The staff finds that it is in its own professional interest to create industry consensus even where none exists.

If a possible industry-consensus position happens to support the position of the economic incumbents—the most powerful and largest members of the trade association—these members in turn will find it in their own interest to cajole or browbeat their colleagues into acceptance of this common position.

The ability of a trade association to articulate a common position on a limited number of issues also provides staff with a portfolio to negotiate deals with other trade associations. These deals are naturally of the so-called log rolling or "I scratch your back, you scratch mine" nature that budget deals by Congress members have. Again, nothing is illegal or immoral about these deals. However, they do open the door to policy changes that undercut market forces and reduce competition.

Deal-Making

Multi-party deals are a challenge to make, so staff at trade associations would not take long to realize that to the extent that it, the staff, negotiates with other trade associations on behalf of its whole industry, without actively bringing members into the negotiation, it's easier to make a deal. Staff will recognize that both corporate (trade association) interest and self-interest reasons support a strategy to cut members out of the loop as much as possible in deal-making: first, it simplifies the potential to make a deal, and second, it enhances the importance that a staff has to its membership. Staff persons can demonstrate through unilateral deal-making on behalf of the association their own indispensable role to support members' needs: a role that could not be undertaken as effectively if the company sent its own direct lobbyists and representatives to Washington.

If trade associations and their staff benefit when they are at the center of the loop, and leave members out as much as possible, it would certainly be in an association's interest and its staff's interest to try to intimidate or even threaten individual members to accept, or at least refrain from vetoing, deals that the trade association has made.

In this theoretical discussion, we see how a system that began as an efficient way for expert professionals to represent the economic interests of individual companies ends up in a process where the self-interest of an individual company is almost irrelevant. The process will naturally seek perceived common interests and will leave individual company self-interest unrepresented and, indeed, unwelcomed.

The reason for the effectiveness of the free market is that it is a bottom-up system in which corporations that seek self-interest interact with each other in a way that, at its best, maximizes everyone's interests. But if the self-interests of companies are aggregated through a trade association process, and if the trade associations bargain with each other to get sufficient political power to pass legislation that achieves their goals, the resultant economy is no longer based on diffuse decision-making from the bottom up and transitions to one based on top-down planning.

Factors for Market Success

With all of the market failures and human failures described above, why is it that free-market economies nevertheless are more innovative than planned economies? What is it about the free-market paradigm that is so consistent with innovation and economic growth (notwithstanding the argument that this connection can be overstated)?

In short, what parts of the economy are less prone to the failures identified in this chapter? How can these failures be corrected?

Interestingly many of the most important market failures and human failures thus far identified apply mainly in the context of a large corporation that pursues a business it has been involved in for a number of years.

On the other hand, it's widely accepted that much of the economic dynamism of the United States economy and the fast-growing economies throughout the world comes from smaller companies and from start-ups. Indeed, many of America's most successful large corporations of this decade were small companies with only a handful of employees just twenty-five years ago.

None of the problems of risk aversion, loss aversion, or status quo bias applies to a small, start-up corporation where there is no corporate value to risk aversion, and no favorable status quo is present to be biased toward. The strength of a market economy is its ability to allow start-up companies to enable individuals to exploit all of the weaknesses and failures in the marketplace, and to "pick up all of the $20 bills lying on the street."

Promoters of the concept of free enterprise most often point to the ability to get rich quick as a sign of the success of a market economy, seemingly ignoring the fact that the existence of such opportunities is itself inconsistent with the model of a market economy that functions properly.

Small companies can have their decision-making focused on one individual to avoid the problems of agency and information transfer because they can have a business model simple enough that one individual can act as a reasonable agent for the entire company. If the company is focused directly on the solution to one of the physical failures in the economy (for example, it seeks to find the 150-percent returns on investment and offers to fix them as a service for someone else for a share of the profits) this company can produce continuous innovations and contribute to dramatic economic growth.

Can Large Corporations Embrace Innovative Policies?

The challenge for economic development is to figure out how to encourage innovation in a larger and more bureaucratic corporation. As noted, this problem becomes much more severe in cases typical of much of the American economy, where a limited number of large corporations shares about 20 percent of the market each and has developed a degree of economic maturity. Environmental policies can work most effectively on this form of business organization.

The types of failures described in this chapter systematically discourage innovation and risk-taking, behaviors that are essential for economic growth in the twenty-first century. While straightforward actions, such as better information on the environmental and energy use consequences of different market choices, might overcome many of the market barriers, most of the market failures and human failures are not easy to correct. And the institutional failures effectuated by the role of trade associations in the political economy are not likely to change soon.

So the clearest way to promote innovation and risk-taking and to achieve their benefits in terms of enhanced growth is to find the areas in the economy where innovation is needed to meet nonbusiness needs and to design market-based policies that achieve these objectives. One of the most obvious areas of nonbusiness need is protecting the environment. The direct benefits of environmental quality are an important value in the lives of individuals, and a clean natural environment assures a more stable and lower-risk business environment.

So a strategy of using environmental objectives as a means of overcoming failures of the market that hold back growth and of enhancing market forces to accomplish both goals should be attractive. Both business leaders and environmentalists should enthusiastically embrace such a plan. This is unlikely to happen unless America can surpass some of the problems of strategic planning and alliance-seeking typical of today.

When large corporations whose employees feel industry-group peer cohesiveness and peer pressure dominate the economy, where their public positions in terms of their private-sector standard-writing activities, advocacy, and lobbying in Congress and Administrative agencies play an important role in the establishment of the rules of competition within the marketplace, it will not be easy to effectuate such changes.

Corporations exert a significant influence on state and national policy in the United States (and in other countries). Many political leaders characterize their decisions on a variety of subjects as "pro-business"; indeed, most anti-environmental policies are defended on the grounds that they are pro-business.

Who decides what truly is pro-business?

On what grounds does business oppose environmental protection?

The answers are addressed in Part 3.

Part 3

The Politics of Environmentalism

Part 1 explores the potential economic benefits that result if we accelerate the use of environmentally preferred technologies and how this process can enhance growth and job creation at the level of trillions of dollars. It also illustrates how the application of such technologies can enhance innovation and produce even greater growth in jobs and economic output.

Part 2 offers information that refutes the prevailing political story that markets already exploit such opportunities, and lays the groundwork to outline how environmental protection policies can enhance free markets, and can tap the growth potential described in Part 1. This section also discusses some severe and generally overlooked failures in real markets related to environmental issues that are likely to be nearly universal in the market.

The question that naturally arises is what can be done? Part 3 introduces some initial steps to answer this question.

Important to note is that a reader who searches for a simple cookbook solution will be disappointed by the conclusions. The problems are complex, and broad and general answers are often incorrect. In short, we need detailed discussions.

However, the problems of failures of markets and of market opportunities for growth are interrelated and it's possible to make some general observations. To speak broadly, the problem is that large-scale failures of markets to invest in innovation are holding back growth and thwarting the achievement of consensus goals, particularly with respect to environmental protection. Broad institutional structures systematically suppress competition even if this is not anyone's explicit goal.

Such failures are unrecognized outside of a few narrow communities, so no effort is expended to solve them, or even to discuss how to solve them, by a broad enough group of stakeholders to be able to come up with better answers and implement them. Worse yet, there is a web of political and policy ideas that demonizes environmentalists in the minds of many in the business and political communities, and business in the eyes of many environmentalists. This demonization is based on myths—convenient but oversimplified stories of what environmental advocates try to do and who they are, or of how corporations work—that are inconsistent with reality.

Such myths described above are fatal barriers to the kind of discussion and deal-making that I believe must be a key basis for solutions. They are also important political barriers that make it possible for lawmakers to vote against the environment even when a majority of registered voters of both parties support environmental protection. So Part 3 begins with a discussion of the myths in Chapters 7 and 8.

Most myths are based on at least an element of truth. Businesses and environmentalists sometimes distrust each other for good reasons, and the political dynamic on environmental law has provided that each side's own weaknesses make the results they are most concerned about more likely to occur (discussed further in Chapter 9).

The purpose of Part 3—the whole book, actually—is to provoke constructive dialogue between those who are interested in environmental protection and those who want to advance economic growth or the interests of business. Not only by myths but also by a more intense refusal to look at the issues is this dialogue impeded. Chapter 10 begins with a discussion of global climate change. At first reading, this would appear to be just one more case study of the kind of results one would expect to find based on the content in Part 1. Yet it is included as a way to ask some fundamental questions about which issues the political system chooses to address and which to ignore.

To ignore a problem does not make it go away. More often than not, it allows the problem to fester. In this chapter I ponder why some key questions about the environment and growth never seem to get addressed by important political and policy constituencies.

Finally, Chapter 11 tackles the recommendations of this book: how should we write well-designed environmental policies to promote economic development?

7 | Myths of the Anti-Environmentalists

Reality can be complicated, so often people rely on simple stories that can explain a complex world in easy-to-understand terms. Frequently these stories or myths have enough connection to the real world that they are helpful and lead to desired behaviors. For example, the myth that in the American economy a person who applies herself, works hard, and stays out of trouble can get ahead and become wealthy has guided many people in the right direction and helped some of them get rich. Even if the story is oversimplified, it serves a useful purpose.

Unfortunately, within the area of the environment, the myths are much less productive. I discuss several apparent myths of the anti-environmental lobby in this chapter, and focus on arguments that the business community has used to oppose environmental policies. (Chapter 8 examines parallel myths of the pro-environmental lobby.)

In both cases these myths not only lead to advocacy positions that undercut the true self-interest of the groups and companies that make them, but also muddy the waters of public policy debate for those who try to make decisions and are lobbied from both sides.

To introduce this subject, a reader who is new to the environmental debate doesn't need to see much anti-environmental and pro-environmental literature on the Internet to realize an important fact: these people simply don't talk to each other. As previous discussions of energy forecasting and economic theory depict, neither formal research nor informal discussions by proponents of one side is referred to in the published work of groups whose opinions or priorities differ. Arguments are not refuted; they are simply ignored.

So the first important myth of anti-environmentalists (and almost in mirror image of environmentalists as well) is that their opinions are developed independently. Instead, *peer pressure* seems to underlie a great deal of the anti-environmental advocacy of the business community. I discuss this in greater detail below.

The Myth of Independent Objective Analysis

Anti-environmentalists take the strong position that their theories and opinions are based on independent and objective analysis that leads them to a correct economic and political opinion. The reality behind this myth is that their theories and actions often are not at all objective or independent but rather quite subjective and based on shared perceived interests.

The nonindependence of the anti-environmentalist position is evident via the close linkages on the Internet between conservative political position statements, business organizations' positions on the environment, and the websites of conservative think tanks. This interconnection is reflected in the ideological uniformity of the responses. It is difficult to find conflicting views among anti-environmentalists: they echo the same types of positions.

Evidently the analysis that underlies anti-environmentalists is developed centrally or at least with central coordination and disseminated in an organized fashion. They have a consistent story to tell, but one that does not reflect the true interests of its proponents as individuals or representatives of a standalone corporation. Instead, it reflects group consensus on the perceived interests of business as an institution or of philosophical conservatives.

To say a central pro-business story characterizes *all* environmental issues is inaccurate: many individual regulatory debates revolve around honest arguments over the environmental risks of a particular action and the costs of change. Myths do not explain everything. Yet a surprising number of the arguments, particularly the broader ones, depend on both sides' myths.

Business people often identify themselves as conservatives, and one of the key, fundamental principles of conservatism is loyalty. Pro-business advocacy seems marked by a strong level of loyalty, a commitment to a type of patriotism that focuses on the advantages of the American system of free enterprise.

This concept of loyalty leads to two sources of peer pressure: pressure to identify as a member of business in general or of a particular industry (and thus to express support for the positions that this industry or the American business community in general support), and negative pressure to come to the aid of ones' allies when the industry is perceived to be under threat.

Group Identification and Support

Numerous forums are available that reinforce group identity both in terms of how to define what positions an industry should embrace or reject and in terms of how to validate a particular set of opinions through social interactions that is limited to those who basically share those positions.

All of the industries with which I have worked have their own internal organizational structure that in practice serves social as well as business and advocacy needs. Most businesses have trade associations that act as a forum for like-minded people to discuss common interests and common needs. Many businesses that work in more technical areas also work with professional societies that have annual or semi-annual or more frequent conventions or conferences at which professionals can discuss matters of common concern, as well as receive new technical information, training, et cetera.

Such business organizations, trade associations, and professional societies are by their nature narrow in focus: they're not the kind of places where people come together to hear non industry-related opinions. So they serve as functions that can consolidate an "industry-wide"view of the world and industry consensus positions on such policies as the environment.

The sort of peer pressure noted above fundamentally and irremediably gets in the way of competition. It creates strong interpersonal values and needs to protect collective industry interest, even at the expense of the interest of the company for which the individual works.

Such a tendency for group cohesion is evident in the dynamics of trade associations. If a trade association represents vigorous competitors within the industry, which in theory should be the case all of the time, the trade association will find that only a few policies exist on which its members can all agree. A larger number of policies will emerge in which a decision in one direction benefits one or two of its members, but the opposite decision benefits other members.

One would therefore expect to see power struggles develop in trade associations, and frequent cases in which one member drops out of an association because it no longer represents his interest. One would expect to see two or-three competing trade associations within the same industry because they represent the needs of different manufacturers or of different manufacturers at varying points in time.

However, the above scenario seldom happens. Over the twenty-five years that I have dealt with trade associations, I have witnessed their leadership's extraordinary efforts to promote group consensus and to retain members. In extreme cases, trade associations have actually threatened individual companies for their positions that have opposed the trade association position.

Peer pressure to develop and advocate a consistent industry-wide position or even a consistent business community-wide position not only makes sense based on how human beings truly react to each other, but also it makes sense as a political strategy. A united industry position would be much more effective with Congress or with a regulatory agency than a divided one. Peer pressure often is used to bring dissenting companies or groups back into line: a group leader can promise that benefits will be dispersed to those who are on the favorable side of a significant issue, and that organizations that are not helpful will be left behind.

The Influence of Interpersonal Relationships

Peer pressure is not always merely psychological and subliminal. Sometimes it is based on the interpersonal relationships between sellers and buyers that are necessary for business success.

For example, consider a potential environmental regulation that would require the use of more steel. One would expect the steel companies to support such a regulation because it increases the demand for their product. But suppose that the major customers of a particular steel company object strongly to the proposed rule. A steel company that sided against its customer might risk retaliation as the customer could change suppliers to a more politically supportive competitor.

Even if the steel company's leaders did not know for a fact that this would happen, the mere risk of customer loss could outweigh the benefits of the regulation. The salience of this risk is magnified by the phenomena of loss aversion and risk aversion. So the steel company keeps silent or even supports its anti-regulatory customer.

This dynamic, to be successful, requires a presumption in the business community that it is normal to oppose government policies and an anomalous and aggressive decision to support them. Otherwise there would be no basis for the customer to assume that it could or should use its market power to limit the policy positions of its partners in purely business deals.

And of course, the use of such power to affect market decisions is directly contrary to all of the assumptions about how free markets work.

The process of business allies coercively influencing the policy positions of their suppliers or customers is not unusual. I have seen this operate in a number of cases, involving both regulations and nonregulatory policies, and across a broad range of companies and industries.

Sometimes the use of peer pressure is an overt strategy to enhance political power. Occasionally, peer pressure is supplemented by direct coercion by groups that represent economic incumbents and that need broader political support in order to succeed. Grover G. Norquist, president of Americans for Tax Reform, an organization that supports conservative causes, was quoted in the *Washington Post*: "If what people remember about you is that you are not helpful, you are probably not going to be first in line when we do the next tax bill."[1]

The importance of peer pressure within the anti-environmental lobby can again be seen by a quick foray into the Internet. The websites of the Republican Party and of several major business advocacy organizations link to a number of conservative think tanks who tell a consistent story about why the opposition to virtually any major environmental initiative is important to protect the economy. The story is based on myths that concern who the environmentalists are, what their true interests are, how the market works, and what the real self-interests of business are—all of which are discussed below.

The Myths about Environmentalists

Many myths exist, which anti-environmentalists truly believe or at least perpetrate, that demonize all those who work for environmental regulation. Here are some of the most egregious examples.

Environmentalists Are Communists

Perhaps the most widespread myth about environmentalists is that environmental activists are truly, at heart, Communists, and that the struggle of the business community against environmental activism is actually a struggle to protect private property from Communism.

To equate environmentalists with Communists is seldom done in so many words by the anti-environmentalist authors, but is quite evident indirectly from their writing, and surprisingly widespread. Chapter 1 notes how a number of advocacy articles on climate change asserted that even environmentalists recognized that global climate change wasn't all that important! Instead, the articles argued, environmentalists recognized that doing something about global warming would require more government control—perhaps to the point of total government control over private industry, as in a Communist country. They asserted that the hidden agenda of environmentalists was not environmental protection, but rather government control of business.

The above speculation is echoed in the *Wall Street Journal*'s citation of business concerns about global climate change: "Many skeptics contend that liberal environmental agendas are behind alarming global-warming headlines, although often skeptics bring agendas of their own."[2] It is difficult to imagine what so-called liberal environmental agendas might mean if not state-run central economic planning.

In at least one case the linkage is made explicitly and publicly. Nationally syndicated columnist George Will wrote in an op-ed column that "For some people, environmentalism is collectivism in drag. Such people use environmental causes and rhetoric not to change the political climate for the purpose of environmental improvement. Rather, for them, changing the society's politics is the end, and environmental policies are mere means to that end." ("Our Fake Drilling Debate," *Washington Post*, 15 December 2005.)

Such a theme recurs constantly on the websites of conservative think-tank organizations, such as the Competitive Enterprise Institute, that are widely cross-referenced on other business and political organization sites. The Institute's president, Fred L. Smith, Jr., wrote a paper available on the Institute's website in 2004 entitled "Eco-Socialism: Threat to Liberty Around the World." In it, he states, "the ecology has become the battleground on which competing visions [of the economy] now engage." And the text states clearly that the two competing visions are his Institute's vision of free markets and the opponents' vision of "centralized collectivist solutions."

This article above is not an isolated case of this story being circulated. The Institute is well linked on the Internet to other conservative sites, and articles and reports of a similar tenor are widespread. Evidently many people believe, or accept the policies of people who believe, that environmentalism is less important in its own right than it is as a "battleground on which competing visions now engage."

The view above would seem to explain much of the business community's knee-jerk advocacy against environmental protection. American business has been concerned about socialism for well over 150 years. In fact, much of businesses' activism against organized labor in the nineteenth century was based on the fear that labor union organizations were the precursor of Communist revolutions that would expropriate property from business.

The fears of Communism are not completely ill founded: many countries, beginning with the Soviet Union, underwent socialist revolutions and did in fact expropriate property. If a real connection between environmental advocacy and Communism did exist, this would be something serious for business to worry about.

Actually, the facts are almost completely opposite: where socialists have expropriated private industry, they have also operated it in a way that is much more irresponsible environmentally than is the case in market-based economies. Centrally planned economies have the world's worst record on environmental protection, pollution, destruction of natural environments, and the most hostility for citizen-based environmental advocacy.

The concerns about government control are equally mistaken. My experience in the Soviet Union, and more recently in China, shows that the level of state control over actual production pro-

cesses in these economies is lower that it is in market economies like America and Europe, often to the extent that even
when the country seriously wants to change business practices
to protect the environment, as is now the case in China, the
level of control necessary to do this well simply doesn't exist.

In other words, a free-market economy subject to the rule of
law is a much more fertile field to implement environmental
policies than a Communist country. Based on the facts, environmentalists should be at least as anti-socialist as businesses.

Yet because of the strength of the myth of environmentalism
as the companion or the stalking horse of Communism, no one
ever discusses these issues.

Environmentalists Will Take Away Our Freedom

A second closely related myth is that environmentalists want
to supplant free choice in the market with government control
over people's lives. This myth also revolves around the issue of
economic fundamentalism.

Economic fundamentalism asserts that government regulations intrude on personal freedom. As shown, it ignores the
question of how private-sector regulations can be equally intrusive but more challenging to change (because the processes to
develop or change them are often so opaque).

This myth of government excess is also inconsistent with many
of the actual questions at issue on the environment. Sometimes
environmental positions require more intrusive government,
sometimes less. Sometimes intrusions into the lives of some
people are expansions of freedom for other people.

If the government sets aside land for habitat protection or
national parks, it enhances the freedom of people to use the
land for recreation or to enjoy the wildlife that will be protected. If the land is private, it intrudes into the perceived rights
of the owner but takes away restrictions on the actions of everyone else.

If the government zones land for two housing units per acre,
it restricts the rights of the property owner and encourages

urban sprawl. Environmentalists generally would support less restrictive government practices in this case.

If the government sets limits on air-pollution emissions, it changes the set of technology options that corporate managers can choose among, and perhaps requires them to upgrade facilities that they hadn't, but does not limit their freedom in any significant way.

If the government sets minimum standards for appliance efficiency, these will not limit economic choices of consumers or producers unless it causes certain features to become infeasible of production. However, in the United States the appliance efficiency standards law prevents the Department of Energy from setting standards that eliminate features. And none of the state efficiency standards has ever done so.

Environmentalists Are Unrealistic Dreamers

Another important myth of anti-environmentalist business people is to characterize their political and ideological opponents as unrealistic dreamers who have no real understanding of business reality.

Business people, not surprisingly, think of themselves as the experts on their business. They often will express knee-jerk opposition to environmental policies because they feel they know their business or industry better than anyone else, and outsiders, whether government bureaucrats or nonprofit organizations, simply don't understand the constraints they are under or the limitations that they face. Environmental policies will, they believe, reduce economic growth because they are designed by nonexperts and opposed by the people who actually understand the industry.

The problem with this argument is that while managers undoubtedly understand the day-to-day operations of their industry better than outsiders could, they often miss the broader issues that occur even within their own plant. For example, when outside consultants find the opportunity for highly profitable pollution prevention or energy-efficiency projects that managers didn't think existed, they show the flaws of this insider-versus-outsider argument.

In some cases, the industry advocate may be unable to understand the difference between group actions and individual actions. That is to say, if an environmental regulation caused the cost of an automobile to rise by $500, a representative from a car company could probably tell you how much sales would decline if the price of his cars went up by that much (and his competition froze prices). But this would be based on the experience of trying to raise prices $500 on the company's product while its competitors were doing something different. The more relevant question—what would happen if the cost of all cars went up by $500?—is one that the industry expert would have little or no experience with because it may not have happened within the time he was on the job.

In economist language, "[business] lobbyists do not know, from their own experience, the general equilibrium context—the adding up of constraints for the economy as a whole—on which they operate. [This] leads them to argue for policies that are not in their own interests, far less those of the public at large."[3]

Environmentalists Are Anti-Technology

Business people often accept mythologies about who environmentalists are and what they are motivated by. Frequently environmentalists are called Luddites or tree huggers: groups opposed to technological progress or individuals whose true goal is a return to simpler, pre-industrial times. And while environmentalists who embrace this view do exist, the major national organizations all consistently support policies that are compatible with and even supportive of a modern industrial economy. The thrust of this book in support of technologies for environmental protection illustrates that the perceived back-to-nature sentiment is truly not much of a threat to business.

The reality is that in a world of six billion people, technology is needed both to protect the environment and to maintain prosperity. Philosophical arguments about whether we were better off when only a million humans lived as hunter-gatherers are irrelevant to policy discussions because it's impossible for the current population to live this way (even if we wanted to).

Environmentalists Are Financially Motivated

Finally, some business advocates believe that environmentalists are "in it for the money." That is, environmentalists stir up anti-business debates simply to raise money and perpetuate their own jobs. But actually, environmental advocacy usually pays much less than comparable jobs in the private sector. The best-paying environmental jobs try (not always successfully) to compete with pay scales in education or government yet remain below those in the private sector.

The Consequences of the Anti-Environmentalist Myths

Most of the proponents of the anti-environmental myths are self-described conservatives, and the main Internet-based sources of anti-environmental mythology are all associated with conservative political positions. But most of the philosophical underpinnings of environmentalism are consistent with and supportive of core conservative values. So as they oppose environmentalism, these anti-environmentalists actually appear to advocate several goals that are the opposite of conservatism, which include the following.

The Loss of Responsibility

Perhaps the key conservative value is responsibility. Conservatives believe that individuals are responsible for their own actions and should receive the rewards of good behavior as well as suffer the consequences of bad behavior.

Yet somehow this concept evaporates when it comes to the environment. Anti-environmental advocacy claims that if a certain chemical is toxic, the victims should pay for their own remediation or suffer the medical damage, but the polluter needn't pay. If a power plant produces fine particles that cause an increase in heart attack deaths and respiratory disease, the producer of the pollution is considered not responsible and the victims are left to their own devices to seek compensation or to suffer.

Conservative attitudes focus on how to protect against potential problems and losses. A key part of responsibility is to assume that everything may not go as planned: the mature individual should prepare. A responsible driver has car insurance, and a responsible homeowner is prepared for both physical and financial disasters.

However, this insistence on responsibility also disappears when it comes to the environment. If climate change poses the risk of disruption to the economy, as well as the environment, the attitude of self-described conservatives has been to ignore the risk and not prepare for the consequences!

Such an attitude is particularly incomprehensible when it comes to questions like health effects of air pollution from power plants. Pollution from electric power stations, primarily from coal-fired plants that were constructed before clean-air regulations, accounts for some 30,000 excess deaths annually in the United States. These deaths are not confined to a particular social or economic class. Investors in the power plants that produced the pollution, and their parents, children, and spouses, are equally at risk for disease and death as people unrelated to the industry. The policymakers at the corporate level who fight pollution controls that would force power plants to clean up are equally exposed to the health risks that they impose as those who they fight against.

An important question to address in regard to the mythologies of anti-environmentalists is: why do these people fail to recognize the consequences of their actions? Such failure to believe seemingly unquestionable and objective facts is one of the most puzzling experiences I have had in advocacy.

Frequently during discussions about how stringent to set a regulation for energy efficiency of a given product, I have observed that a decision to set the standard at a more stringent level might save, for example, 500 lives per year. Therefore, the cost or difficulty to achieve the higher, more ambitious goal should be balanced against the value of 500 saved lives.

Ordinarily you would expect such a prospect to save or lose 500 lives to motivate people regardless of their political position, including what their eventual advocacy position was on the regulation. In fact, the response I get to these observations is uncomfortable laughter: the listeners are in denial. The industry representatives simply refuse to acknowledge emotionally, if not intellectually, the occurrence of these deaths. They neither attempt to refute the data nor deny the connection: they simply dismiss it.

The mythology that bad effects can be willed away if we simply ignore them is an important and unacknowledged stream of anti-environmental advocacy.

Another odd example of this problem is the issue of mercury in fish. It is increasingly well understood that fish, particularly large long-lived pelagic fish like tuna, are contaminated with mercury, most of which is caused by pollution. Excessive fish consumption exposes people to health-threatening levels of mercury poisoning.

Obviously, contamination is a threat to the tuna industry. It would be logical to expect that the tuna companies would work in alliance with environmentalists to require reductions in mercury pollution that would translate directly into improvements in the safety of their product. In fact, the U.S. tuna industry is silent on this issue, despite years-long efforts by environmental organizations to forge an alliance.

The Policy of Denial

Denial is part of a broader myth of the anti-environmentalists. Many anti-environmentalists set forth the myth that environmentalists are Chicken Littles who exaggerate or make up threats, whiners, non-team players, or even anti-American, because they point out the adverse consequences of proceeding down current paths.

A surprisingly large component of anti-environmentalist literature is articles and books that refute alleged environmentalist claims that environmental quality is getting worse, or that postulate that economic development automatically improves environmental health so that explicit environmental policies are less important than economic growth-inducing policies.

A component of this argument is that people who point out the flaws with the current industrial system are disloyal or un-American. Apparently the tenet of this myth is that if one is optimistic, good things will happen. Its followers apparently believe that simply doing nothing about the environment and having faith that things will get better will solve all the problems.

The issue of whether the environment is cleaner or dirtier is a spurious issue, which makes this element of mythology particularly distressing. Analysts who argue that it is important to determine whether environmentalists predict that things will get better or worse miss an important distinction between a forecast and a plan.

The distinction between forecasting and planning is best illustrated by the following story: late one night a small, meek, middle-aged bank examiner looks over the accounts of his multi-billion-dollar organization and unexpectedly realizes the bank is on track to lose $2 billion over the next financial quarter. After many sleepless nights, he builds up his courage and walks into the company CEO's office and points out his findings. The CEO looks at the information and realizes that the examiner is right. The CEO orders a major reorganization of the bank's business plan. As a result of this aggressive management action, the loss is averted and the bank breaks even for the quarter.

The bank examiner has watched all this happen and is surprised that neither a promotion nor a bonus is offered. So again he pumps up his courage and walks into the CEO's office and asks what happened. He reminds the CEO that he had predicted that the bank was headed toward a $2 billion loss and as a logical result of that prediction, the CEO changed course and prevented the loss. Shouldn't he receive some recognition and reward?

The CEO spoke angrily: "You don't deserve any reward at all. You made the prediction that the bank would lose $2 billion this quarter. You were wrong, it broke even."

The bank examiner was so outraged by this that he had the nerve to respond: "But what if I had predicted that the bank would have broken even this quarter?"

The CEO answered with similar arrogance: "You still would have been wrong." He dismissed the bank manager without recognition.

The story illustrates how planning or policymaking is a completely different endeavor than forecasting. The bank examiner was engaged in planning. What he meant to convey was, if the bank proceeds and does not make a change, it will lose $2 billion. That statement would have been correct. Instead, the projection was expressed (or as the CEO understood it) as a forecast: the bank is going to lose $2 billion unconditionally.

Predictions of environmental disaster are exactly the same. Environmentalists who present credible disaster scenarios warn that unless something changes, a number of frightening consequences will ensue. Usually world political leaders have been responsive enough that when presented with strong evidence that business-as-usual leads to environmental disaster, business-as-usual changes.

So to "predict" an environmental disaster, environmental advocates help to prevent it. If they succeed, then the so-called prediction will be wrong. Yet perhaps it was necessary to assure that the environment becomes more healthful.

Thus the issue of whether the environment improves or degrades in general is meaningless: what is important are the predictions of what the consequences of particular changes in environmental policy might be.

These issues are lost on the myth-spinners of the anti-environmental movement, who seem to see cornucopia proceeding from an unfettered continuation of the status quo.

Perhaps this myth is based on the real-life observation that in the business world, a positive, optimistic attitude often leads to a better outcome. Clearly if all business leaders believe that the economy is strong, then they will prepare accordingly; they will order new equipment and hire new workers. These investments themselves will make the economy stronger, at least in the short run.[4]

A similar principle holds true for an individual company as well. If everyone thinks a certain corporation will do well, its stock price will go up, as will its credit rating, and it will in fact do better.

Unfortunately this line of thought is unsuccessful outside the business world. The best example is health care. If a person is overly optimistic about her health, she will ignore danger signs and seek medical advice too late. If she does this long enough, she risks the result that the problem may become untreatable.

Environmental problems are like health problems: positive thinking won't resolve them—it may actually make them worse. (There is no placebo effect for the environment.) Perhaps this particular myth—the importance of optimism—illustrates the relevance of group identity, peer pressure, and loyalty to the mythology structure of anti-environmentalists: people who criticize the status quo are not part of the team, so to speak, and are thus not credible as individuals, regardless how truthful their words.

The Power of Deception or Self-Deception

The specific issue of loyalty, which defines some sources as inherently credible and others as inherently noncredible, leads to a particularly troublesome aspect of anti-environmental mythology. In many cases it crosses over the line from philosophy and ideology to deception (or perhaps self-deception): in many cases, the anti-environmentalists simply get the facts wrong.

I find it somewhat difficult as a nonmember of this community to understand the roots of the problem of consistent factual error, because almost none of the anti-environmental literature claims that the environment is unimportant. A plausible argument can be made about any environmental policy that, if one looks at the facts, the pain might outweigh the gain. Yet if this is the core belief, then why deny the facts, rather than make the direct benefit-cost argument?

Perhaps the most evident example of deceptive advocacy recently was in the context of how to regulate mercury emissions. Recent research in the 2000s has demonstrated the health risks of elevated levels of mercury in fish, most of which is anthropogenic in origin, and many environmentalists and government agencies have warned the public against heavy consumption of fish as a safeguard against the ingestion of too much toxic mercury. Yet an editorial in the *Wall Street Journal*, citing a conservative think tank, argued that no health risk is associated with high mercury levels in fish and claimed that the evidence for human risk was minimal.[5]

The editorial presents a startling and distressing argument to see in one of the nation's leading newspapers: mercury has been known to represent a toxic threat for well over 100 years. No possible scientific question exists about its adverse effects on humans, and there can be little question that frequent consumption of mercury-contaminated fish leads to an increased risk of mercury poisoning.[6]

What sort of thought process leads the editorial writer to publicize such nonsense? What process leads a mainstream newspaper to accept uncritically such obviously mistaken claims?

Unfortunately such examples occur with some regularity. Anti-environmentalists have attempted to deny many of the most scientifically evident issues, whether they relate to the deaths caused by air pollution or the decline of economically significant species (like cod or squid or salmon).

Perhaps the most consequential case of denial is how the George W. Bush Administration continues to refuse to accept the scientific reality of global climate change to a degree sufficient to cause it to take any actions to mitigate the greenhouse gas emissions that are responsible for the problem. The issue is not merely the Administration's position, but the refusal of the business community (when it speaks with a single voice[7]) and of most of the Republican Party[8] to challenge this position, despite strong self-interest in doing so.

Little serious debate remains in the scientific community about the reality of the threat. If industrial emissions of green-house gases continue to grow, as nearly all economic models predict, then the earth will heat up by several degrees by mid-century. The consequences of this warming trend are likely to include more frequent and more severe storms, and may include outbreaks of disease or crop failures on a massive scale. Because weather details are impossible to predict, even more severe but unforeseen consequences may be imminent.

It is impossible to insure against such risks—the entire insurance industry is not big enough to handle the size of the liability. We saw in 2005 that a single hurricane could cause damage estimated well in excess of $100 billion. What would be the consequences of repeated storms?

One would expect that a conservative attitude would be to use policies to mitigate the risk and to spend a little money now to reduce the risk of substantial payments later. Furthermore, it should be expected that a careful analysis of actual costs should underlie business's and conservatives' positions. Yet this hasn't happened.

Recall in Chapters 1–3 that the actual costs of serious emissions reductions are negative: cutting emissions is not merely a cheap lunch, but a lunch they pay you to eat. And more comprehensive analyses of what it would take to meet the international standard for climate protection—the Kyoto Protocol—show that all sectors of the economy could benefit.

These analyses, and the story of their denial, is discussed in Chapter 10.

One could perhaps understand these distortions as advocacy techniques to benefit economic incumbents who perceive (perhaps incorrectly) that they benefit from the status quo, but it is surprising that the business community and the conservative political community has not challenged these examples of junk science or downright misrepresentation.

Myths are a common way to organize and simplify a complex reality. The myths described in this chapter, like most myths, are not total fantasy, but have some basis in truth. Some environmental activists worry about risks that have no scientific basis. A few environmentalists probably do support Communism. Yet reliance on these myths as a guide to corporate strategy or as a model for policy decision-making would be fundamentally wrong.

Not only are myths incorrect, but also maladaptive; they fail to lead their believers into directions that will address true business needs.

The mythological distortions of this adversarial dynamic between anti-environmentalists and environmentalists are not limited to one side only, as I discuss in Chapter 8, which explores myths held by the environmentalists themselves.

8 | Myths of the Environmentalists

Environmental advocates and their organizations are more diverse ideologically compared to anti-environmentalists, or at least express a greater diversity of public positions. While some public information is shared broadly among different environmental organizations, particularly the large national or international advocacy organizations, a number of different philosophies and opinions exist concerning environmental protection. And since loyalty is less of an issue in the environmental movement splits over political positions are common.

Nevertheless, some general myths emerge in which the public writings of environmentalists tend to be similar. These myths are discussed in this chapter.

I believe that these myths are equally counter-productive as those of the anti-environmentalists. In many cases, they are close to mirror images: oversimplified attitudes by environmentalists provoke equally oversimplified reactions by business people or conservatives, or vice versa, and both of these attitudes hinder constructive engagement and dialogue. The following are some examples.

The Greedy Corporation Myth

The first myth is to meld environmental issues with other progressive causes. Environmental literature is full of stories that claim that greedy corporations pollute the environment, and harm people's health "just to increase profits."

While the direct purpose of such a statement could be explained easily by saying that corporations ought to make profits from actions that don't harm people, a simple reading of such a statement implies that the environmentalists believe something is inherently wrong with profits, or perhaps even with corporations. The reader can infer that the environmentalists believe that all corporations are greedy, or that corporations by their very nature will ignore human misery if it enhances profit.

Greed is different than profit maximization, but this myth glosses over such a distinction. A profit-maximizing corporation will follow environmental laws, and may find that making decisions in a way that respects environmental quality is good business. Many businesses act in this fashion. Profit-maximizing firms often believe in corporate social responsibility—in many cases because they believe that it is good business practice.

A greedy corporation is one that takes actions that increase its profits unfairly or at the expense of others. Environmentalists should be concerned if a company ignores its pollution or other environmental effects and selfishly causes damage to others. This behavior does occur in some instances, but it is not the root cause of most environmental problems.

If a corporation acts greedily, it may also try to increase its prices unfairly. This may or may not be a big problem in the real world, but even to the extent it is true, it is not an environmental problem.

Ironically the type of business behavior that harms the environment often is a failure to maximize profits, as shown in Part 1. Many environmental problems are the result of corporations that are insufficiently motivated by profit, not overly profit-minded.

The greedy corporation myth has a number of problems. First, it suggests that some other form of human organization might be better. If corporations are greedy, would municipally owned organizations be better? Would government agencies do a more public-spirited job of protecting the environment? This question illustrates how the environmentalist myth feeds the flames of several of the anti-environment myths.

The suggestion that some other form of organization would be more sensitive to environmental issues is incorrect. There does not seem to be any evidence that corporations are any worse polluters than anyone else with the same level of economic responsibility. In some areas, such as electric or water utilities, transportation providers, package delivery services, and others, we can make direct comparisons, and the corporations behave no differently than the other institutions. We can compare municipally owned electric utilities with their investor-owned counterparts and find examples of environmental leaders in both camps and in about equal proportions. Similarly the anti-environmental utilities include both publicly owned and privately owned companies, again in about equal proportions.

Just as some people are greedy, some corporations will be as well; but the same influences will lead nonprofits or government agencies or anyone else to value their own well-being unreasonably over that of others. However, most of the examples in this book show that to the extent that corporations are responsible for environmental problems, it is more an issue of control, or laziness, than it is of greed.

Businesses cause far more environmental damage when they join forces with their competitors to lobby against environmental protection policies than they do by polluting directly up to the limits of the law, or in violation of the law. Much of this lobbying serves the protection of economic incumbency, which is an issue of control (so that economic incumbents can restrict potential competitors) or laziness (so that they do not have to innovate), both of which are the opposite of greed because they fail to increase profits. (Instead they safeguard the ability to earn about the same level of profits but with less risk.)

In opposition to the myth of greed, some corporations make significant financial contributions or lend staff time to civic or charitable causes, and encourage their employees to contribute or volunteer. Many corporations have adopted principles of social responsibility, which include environmental responsibility, and produce annual reports that document their progress. And, as noted previously, some companies have made outstanding progress on waste reduction, recycling, or energy efficiency.

Why is the myth of corporate greed used so heavily by environmental organizations?

This type of accusation taps into broad emotions that have supported a number of other progressive political causes. Populism beginning before the 1890s opposed corporations just because they were corporations. Populism often drew on stories in which small farmers, modest hard-working families, were forced to deal unfairly against big corporations located far away. These outside corporate interests were unfeeling and imperious, and profited at the expense of the common person.

A more modern version of this myth allies environmentalists with consumer advocates. Again, the ideology is anti-corporate in general and often assumes that corporations inherently will try to steal money from consumers whenever they have the power to do so.

Many environmental policies actually do benefit consumers, so this alliance is not without sound foundation. Yet these same policies also benefit businesses, and for the same reasons, so we would expect to find just as strong a business/environment coalition. But we don't. Part of the reason is that environmentalists do not view corporations as natural allies, and do not try to court businesses as strongly as they do consumer organizations. It is easier for environmentalists to fall victim to the myth of the evil, greedy corporation.

Many environmental advocates have tried to make political alliances with labor organizations. On one level, this is not surprising because as we have discussed, environmental protection policies tend to produce more jobs (although this probably is not the reason for the attempted alliance). More likely, the reason resides in the political realm: labor is a natural ally if your perceived adversary is big business. The reason may also be to try to align environmentalists with other organizations that protect the interests of the average person—in this case the worker. Yet again, this supposedly pro-worker position is also an anti-management position. Certainly business would perceive it this way.

Also, environmental organizations often perceive a natural alliance with low-income and in particular minority advocacy organizations. A factual basis for such an alliance does exist because numerous studies have shown that poor people and minorities are subject to disproportionately high health risk from environmental hazards and suffer the worst from environmental disamenities. The term "environmental justice" makes sense on a lot of substantive grounds. However, like the other alliances, it creates troublesome political connections.

The political dimension is that virtually all of the allies that environmentalists appeal to—populists, consumers, labor, minority and low-income communities—are part of the coalition that has been the core support for the Democrats since at least the Roosevelt era. And so if environmentalists see themselves as supporting the interests that elect Democrats, it follows that Republicans and Republican-oriented businesses could invent a mythology that paints environmentalists as "the other" and, therefore, find them threatening.

In its extreme form, the association of environmentalism with other identified progressive causes could even spawn the fear of Communism that is one of the core myths of anti-environmentalists. This association is natural because the one thing that all of the supposed allies of environmentalists have in common is an economic interest that pits them against established wealth in general and established private enterprise in particular.

Individual corporations can be greedy or they can be socially responsible. Or their actions can simultaneously reflect some of both characteristics. In the real world, most things are neither all black nor all white. The myth of the greedy corporation overlooks this basic principle, which is also related to the next myth: that so-called bad people pollute the environment.

The "Bad People" Myth

A second troublesome myth of the environmentalist community is to characterize problems as to the fault of "bad people" rather than bad policies. Particularly in fund-raising literature, environmentalists often single out individual companies or persons as blameworthy, and they seem to imply that if different people were in charge, the results would be better. Thus, environmental alerts commonly make such statements as, "Jane Doe has polluted this beach and that's why you can't swim there," or "the people who run All-for-Profit Company are responsible for the pollution that causes asthma in innocent children."

While the above statements embodied in this myth may literally be true, the myth is troublesome because it creates the false impression that individual bad decisions are responsible for the problem. Many business leaders who have been similarly criticized take these accusations personally and find them quite painful. Unfortunately for the environment, such pain steers them away from a desire to do better and instead fosters anger at the accusers.

And in many cases the anti-environmental actions are a consequence of much broader influences in the marketplace. For example, if General Motors (GM) sells many polluting and dangerous SUVs, the reason may be that market barriers, market failures, and human failures prevent the introduction of even more profitable and preferred technologies. GM, the corporation, might be to blame (I think it is) because it joined other manufacturers to lobby against policies that would encourage better fuel economy, lower emissions, and better highway safety, but given the current business environment, GM employees are not to blame because they produce and sell what they do.

The distinction above is important because the "blame the bad guys" philosophy fuels the fires of business's fear of oppressive government control and strengthens the anti-environmentalists myths.

Criticizing Collaboration

The degree of criticism of the bad guys can be fairly extreme. In my work with the Natural Resources Defense Council (NRDC) to promote energy efficiency, we have often entered into partnerships and worked in cooperative ways with utilities that run energy-efficiency programs. NRDC also works cooperatively with utilities to obtain a regulatory structure that allows a utility to increase its profits to the extent that it improves its customers' energy efficiency in a cost-effective way.

Yet some environmental organizations have criticized NRDC of "selling out" because we have worked in partnership with utilities and defended them when their pro-environmental commitments were challenged or when their programs were under attack. The issue with these organizations was that the utilities were so thoroughly perceived as the bad guys that no matter how thoroughly they reformed their behavior—trying to sell less rather than more electricity, for example—they were to be blamed.

Even worse, environmentalists who worked with them on these endeavors became part of the "bad guys" as well, "selling out" to the "corporate polluters." (As a note, NRDC as a matter of policy does not accept money from investor-owned utilities to avoid the reality as well as the perception of selling out.)

The Corporations Are Responsible for all Environmental Damage

A variant of the myth of bad people is a tendency for environmentalists to be anti-corporate in general. Corporations are often accused of being the primary cause of pollution or other environmental problems. In truth, many corporations do cause serious environmental problems; yet frequently noncorporate entities are just as bad.

One area where we can compare corporations and noncorporate entities in terms of their pollution emissions and their political advocacy is in the electric utility industry. Many electric utilities are municipally owned and a few are agencies of the federal government. A relevant question to ask is how they differ in their treatment of the environment.

In my experience they do not differ in any systematic way. There are many investor-owned utilities that are strong advocates of environmental protection and that strive to improve their own environmental performance in meaningful ways. But there are also some that will defend their right to pollute aggressively, and there are some that fall in between. The most environmentally conscious municipal utilities and federal agencies perform as equally well as the best private corporations, neither better nor worse, and the least environmentally oriented publicly owned utilities are just as insensitive to environment performance as their investor-owned peers.

Aside from corporations, individual actions—or inattention—cause many other environmental problems. Urban sprawl in the United States is largely a consequence of individual decisions, not corporate choices. (Of course government policies that subsidize sprawl and private-sector informal regulations that discourage smart growth enable these decisions, but these are separate from individual corporate decisions.)

In developing countries individual villagers create many of the threats to the survival of wildlife in their encroachment onto habitats; and the household need for firewood to cook or heat in such areas contributes significantly to deforestation. (This observation is not intended to blame the villagers: there are many reasons for their behavior patterns that are entirely beyond their control.) Corporations are far less a contributing source to these problems.

The "Small Is Beautiful" Myth

A third myth of the environmental community is that "small is beautiful," that decentralized, family scale, small systems, organizations, and communities are inherently good for the environment. This idea is categorized as a myth because little evidence exists to support it, and some evidence refutes it, but mainly because it's another example of environmentalists over-reaching in an attempt to address issues that have little to do with the environment.

My point is not to judge whether the concept that small is beautiful is correct, but rather to point out that it is incorrect to identify it as an environmental issue.

For example, households that burn wood or coal are a major source of air pollution in regions where regulations do not address these problems. Individual use of wood or coal to heat a home is an inherent problem of smallness: to burn coal in a centralized power plant would cause less pollution (although still not the best choice environmentally or economically speaking) because environmental controls can be installed and maintained at higher levels of performance in a power plant than in a home.

Substantively, not much benefit of smallness can be demonstrated from a purely environmental point of view. Large corporations pollute a lot because they are big; yet per unit of production, small companies often pollute substantially more than larger corporations. Often small producers pollute more because they lack the technical skill, managerial commitment, or the capital to use advanced pollution-minimizing equipment and processes. And small rural communities, as opposed to compact big cities with high quality public transit, consume far more gasoline and auto-related, environmentally troublesome materials per capita.

Clearly some examples show how smaller scales of organization work more in harmony with the environment than larger ones, but other examples where the reverse is true are equally present. To be sure, not much of the literature of the environmental advocacy community deals with the small-versus-large-scale issue. Yet to the extent this information is presented, not only can it lead to misleading environmental choices, but also it reinforces the us-versus-them anti–corporate mentality that further distances environmental advocates from business, and is responsible for much of the lack of progress.

Back to the Good Old Days

A related kind of story that is often found in environmental literature is the nostalgic belief that things were better in the old days, and that economic progress causes environmental problems.

The truth is much more complicated than that. In colonial days, an American family used about 250 MBtu of energy annually, mainly from wood burning. Wood smoke is full of toxic chemicals, and particularly when combusted in open fires and fireplaces, wood produces prodigious amounts of pollution. Today an average American family consumes far less energy in its household usage and, unless its electricity is derived from non-pollution-controlled coal, also produces dramatically less air pollution.

Many environmental problems, such as widespread dispersal of toxic metals like lead, were common in the Roman Empire. Deforestation was a major issue on the Greek peninsula over 2,000 years ago. To the extent that the problems were less severe back then, it was only because there were fewer people around.

Another problem with the so-called old-days-were-better myth is that it feeds the anti-environmental myth that environmentalists are anti-technology or Chicken Littles. It further exacerbates the view of the environment in which increasing environmental protection comes at the expense of the interests of the business community.

The myths of the environmentalists are as maladaptive and mistaken as those of business. They have not led to strong and effective alliances with potential supporters in labor, while environmentalists' successful joint advocacy efforts with other members of the Roosevelt coalition have been based almost entirely on pragmatic recognition of mutual interest on a particular issue rather than ideological partnership.

Yet these myths have successfully angered businesspeople and presented barriers to dialogue, and in many cases to assembling issue-specific coalitions with businesses to support common objectives.

The myths described in the last two chapters are based on faulty oversimplifications of what the real interests and concerns are for both sides. In the next chapter, I attempt to identify those concerns, and describe the nonconstructive ways in which advocates of both interests have interacted. As we identify the real problems we may begin to replace myths that focus on the demonization of the adversary with dispassionate discussions of differences in interests and actions that can be discussed and resolved.

9 Legitimate Concerns of Business and Environmental Interests

Up to this point I have focused on the importance of regulation as underpinning free markets rather than a force in opposition to markets, and on reasons why business opposition to environmental regulation is illogical. I have also shown how business opposition to regulation is often just a way to avoid the real issue of business opposition to changes in the status quo.

Yet notwithstanding the aforementioned problems, real reasons indicate why industry should oppose many specific environmental regulatory initiatives based on valid self-interest. Many of these objections could be eliminated or at least mitigated if a better atmosphere of dialogue and collaboration existed between business and environmental advocates, and more broadly between business and government. This chapter explores some of the problems of the myths described in Chapters 7 and 8 and shows how these myths create a vicious circle that makes resolving them more difficult.

Business's Concerns about New Regulation

A business that faces the potential for new regulations over the environment (or for other reasons) could have several types of legitimate concerns. Among the most significant of these are:

- Intrusiveness. The regulations require lots of paperwork or inspections to demonstrate compliance, or the presence of government inspectors that waste managers' time.

- Inflexibility. The regulations may require industry to do one particular thing when an alternative approach would work better and cost less.
- Cost to demonstrate compliance. In many cases, the industry must perform tests or keep records or databases to show that they are in compliance; these procedures can be costly, cause frustration, and can divert management attention from its primary business activities.

These are real issues that are important, and it warrants the effort to see how these problems can be mitigated by better design of environmental policies.

Intrusiveness

Intrusiveness is an important problem because many companies operate in ways that are not obvious to outsiders and competitors and that confer advantages on that particular company. If government inspectors are allowed too much access to their operations, there is a risk that competitors can learn about their management processes, technologies and product formulas, business model, or cost structure and take advantage of that knowledge.

Even when issues of proprietary information are absent, the need to keep records and to interact with government officials on environmental issues can be an annoyance and a source of expense.

And even if the actual time spent on such compliance is unremarkable the need for management attention is an issue. Whenever a regulatory requirement exists, someone must take responsibility for how to deal with it—to answer the letters or phone calls and to compile and file an appropriate response. And a senior manager needs to take responsibility to ensure all tasks are completed.

Inflexibility

The issue of inflexibility has been written about extensively. The basic problem for a business is illustrated in the following example: an environmental regulation requires a given invest-ment to be made in cleanup although the plant manager real-izes that a much smaller investment in something not covered by the regulation could achieve the same or even greater emis-sions reductions at much lower cost.

Because of inflexibility, approaches like "cap and trade" or so-called pollution bubbles have been developed, in which reduced emissions from one source can be traded off against a failure to reduce emissions from another source. Energy performance standards for new buildings is an example of a more flexible regulation.

Sometimes the problem of inflexibility is bureaucratic; other times the government officials who promulgate the regulations lack the proverbial big picture approach on how to achieve their objectives with the least cost.

Yet sometimes the perceived solutions to allow tradeoffs are difficult to implement. Some sources of pollution are difficult to measure, or challenging to audit the accuracy of measurement. In such cases, a more flexible system may be more expensive, or may undercut the goals of pollution control. An obvious dif-ficulty is if the biggest, cheapest reductions are most difficult to measure. In this case the government enforcement official will be concerned that the plant implemented measures that cut emissions on paper but not in reality, while trading away the requirement to do procedures that are proven effective.

A real issue exists about the potential tradeoff between flexibility and effectiveness of a regulation, as well as deeper questions over whether flexibility can induce the same level of innovation and economic growth potential as more directed regulations. These will be addressed in Chapter 11.

Cost

The cost of regulatory compliance is an issue particularly for small business, whose operations may not have developed accounting or oversight techniques to make the process of reg-ularly filling out one more set of forms straightforward. Business often complains, apparently with some justification, about government agencies that pay insufficient attention to the burden that demonstration of compliance places on corporations.

In addition, recall earlier how in many cases business is not opposed to regulation, per se, but rather objects to government efforts to write the regulation as opposed to industry efforts. While advantages to government-sponsored regulation are possible—the ability to have the regulation written by objective and disinterested parties, the ability for broad public participation in order to get the right answer, and finally the ability for litigation to check the abuses in the system—some weaknesses with government-sponsored regulation, particularly from a business point of view, are also evident

Government officials who write the regulations may be disinterested, but they may also have less expertise about how the regulated industry works than the corporate staff members who participate in the regulatory proceeding. A risk always exists that the government official, who does not understand how business actually works, will impose a nonsensical decision on an industry. This is rarely a problem if the regulation is written within a private industry context.

Chapter 7 presented myths of anti-environmentalists and provided illustrations of a number of specious reasons why businesses should not trust environmentalists to develop good public policy. More realistic reasons why business might not trust environmentalists are presented below.

Reasons Why Business Distrusts Environmentalists

First, business may have concerns about the technical soundness of the environmentalists' point of view or even about its reasonableness. This is actually a bilateral concern, but that doesn't make it any less real. Environmentalists, by the very nature of the profession, will know less about how businesses operate than staff or consultants to the corporations. The very breadth of material that environmentalist staff has to master makes it likely that it will operate at the limits of its knowledge in a number of areas where it is dealing with business.

Lack of Standards

A related problem is the lack of credentialing for environmental organizations or their staff. While some national organizations employ recognized technical experts, it is difficult or impossible for business to distinguish between a professional environmental organization and one that relies on volunteers, or between an organization that chooses its staff based on technical/economic expertise or simply the ability to demonstrate concern and work hard.

More broadly, no standards are available on who can self-identify as an environmental organization. A group of neighbors who oppose a particular project in their own area can describe themselves as environmentalists notwithstanding the fact that the project benefits the environment and that established national environmental organizations might disagree with their position. Business is not likely to make the distinction.

A hypothetical example of such a case would be a neighborhood organization that opposes a new transit station in its neighborhood because of concerns that it might bring crime or parking problems. An environmental organization with broader knowledge and perspectives would note that the benefits of the transit station to improve location efficiency and reduce driving in the neighborhood are likely to outweigh any issues of negative neighborhood impact, and may be able to show that the neighborhood effects are mostly favorable. But this opportunity for dialogue may never be realized.

Anti-Business Bias

Business also tends to view environmentalists as irrationally prejudiced against business and growth. Certainly there are public-interest organizations whose primary motivations are anti-growth and anti-business. Much of the fund-raising literature of even the environmental organizations that frequently work with business will frequently contain material that can be correctly perceived as biased against business. Given the diversity of such public-interest organizations, it is difficult for a business to grasp how it might begin to establish better working relationships with respect to environmental advocates.

It is not easy to work in partnership with organizations who throw verbal stones at one's organization. In my environmental advocacy I have had to spend considerable effort both to restrain my harsher-tongued colleagues and to reassure negotiating partners that ours was a reliable organization with which to work.

Environmentalists' Concerns About Business

Environmentalists have many parallel concerns in dealing with business. Unfortunately, as we will learn, many of these mutual concerns lead to a vicious circle in which both sides' responses to distrust by the other leads to the creation of genuine reasons for future distrust. Here are some of the issues.

Ethics

One concern of environmentalists about business is that certain businesses truly do not care about the environment, but instead care only about maximizing their own private profits. Even if the environmentalists agree that nothing is wrong (actually something is right) with increasing profits, a common perception is that business will act to maximize profits even when doing so has direct negative consequences on other people.

This sort of behavior, which in my experience is atypical of most companies, is one of the sources of the environmentalists' myth of the greedy corporation described in the last chapter. Most people in America would probably agree that the primary purpose of a corporation is to make money. Yet there are ethical limits on how business should be able to make this money.

It's expected that if an airline discovers a safety problem with one of its aircraft just before take-off it will postpone the flight in order to fix the problem or cancel the flight if the problem cannot be resolved, even if they lose money as a result of this decision. Likewise we expect a food manufacturer that discovers contamination in one of its products to recall the whole lot before people get sick.

Similarly we would expect that if a company discovers that its pollution kills people that it would attempt to abate it. However, this occasionally fails to happen, and when that's the case, the action breeds mistrust.

This issue of distrust is sometimes coupled to a belief that profitability is a zero-sum game—meaning that more profit to a company that pollutes means less money for everyone else.

This issue above is more than just a reflection of the myth of the greedy corporation; the real problem I have observed is less a matter of greed as it is resistance to change. A greedy, but well-managed, corporation would see the business opportunity in environmental policies, or at least in some of them, and would be aligned with environmentalists as often or more so than it would be opposed. A corporation that takes the consequences of its actions seriously—that is, one guided by ethical values—would at least look at how more socially responsible behavior could be in its self-interest.

Some environmentalists are also concerned about the issue of big business in general as an anti-democratic institution. Certainly the extent to which business lobbying can lead to the promulgation of laws or regulations that are contrary to the public interest is such a concern.

Broken Agreements

There are practical reasons why many environmentalists do not trust big business. Perhaps foremost of these is the concern that some businesses opportunistically break agreements that previously had been made.

My experience shows two important examples of such untrustworthiness and also offers a reason why it may occur. In

Chapter 2, I discussed the negotiations in 1986 that led to the promulgation of the National Appliance Energy Conservation Act. As part of the negotiations, the involved parties—consumers, environmentalists, states, and energy-efficiency advocates on one hand and manufacturers on the other (with a number of other stakeholder organizations involved)—wanted to resolve the issue of whether appliance efficiency standards would be promulgated on a state-by-state basis or on a federal basis. The agreement that resulted was intended to be a permanent solution: the Department of Energy (DOE) would conduct an ongoing program to improve appliance efficiency regulations at the federal level and states would largely be preempted from future activity.

A commitment surfaced among the main participants that the negotiated solution would be permanent: that one side would forgo unilateral changes in the basic appliance efficiency law, but instead would work in consensus with the other organizations.

Yet only nine years later, when Congress became increasingly conservative, some of the same stakeholders on the corporate side lobbied Congress for fundamental changes in the appliance efficiency law. A one-year moratorium on the issuance of appliance efficiency standards resulted from an extensive political battle; a moratorium that set the federal program back several years. And it could have been worse if the more aggressive companies had prevailed, or if some companies had not been outspoken in defense of the original agreement.

Clearly industry would have been appalled had the factual situation been the reverse. If the environmentalists had taken advantage of a suddenly leftward shift in Congress to overturn state preemption and allow both state and federal programs, industry would have considered that treacherous behavior.

A second example from the appliance efficiency area is the negotiated refrigerator standards of 1994. As discussed in Chapter 2, pro-efficiency interests and the refrigerator industry reached agreement in 1994 on a joint recommendation to

DOE for the 1997 refrigerator standard. The recommendation was submitted to DOE in 1994, but the Department was slow to respond. The Final Proposed Rule was issued almost a year later. At that point, several of the companies reneged on the agreement; Whirlpool was the only company that had continued its commitment to support the negotiated standard.

Whirlpool apparently had some strong self-interest reasons to stick with the deal. The company had made significant investments in components that would allow it to comply on time. If the standard were delayed, these investments would be unproductive for the period of the delay. While eventually DOE did promulgate the originally agreed-upon rule, the effective date was delayed about three years. The environmentalist side was distressed that industry felt it could walk out on an agreement so quickly after it was made simply because it was politically possible.

Perhaps it wasn't so much that the corporation broke its word, but rather that the agent who decided to take advantage of the new political situation wasn't aware of the extent to which another agent within the same company had committed that company to the previous agreed-upon path. But one of the lessons that environmentalists drew from these bad experiences was that the corporate commitment to the agreed-upon deal apparently was not, in fact, corporate-wide.

In response to this experience, environmentalists continued to negotiate with industry on consensus standards, but insisted that officers of the board of directors rather than corporate staff sign the final agreements. So far that seems to have worked: consensus standards on ballasts for fluorescent lamps and clothes washers negotiated in 2000 are being implemented as planned with effectiveness dates in 2004, 2005, 2007, and 2010; and over a dozen new consensus standards were negotiated in the succeeding five years and were enacted into law in the Energy Policy Act of 2005.

Circumventing Regulations

Viewed from either side, it is evident that the political dynamic leads to a number of vicious circles. Inflexibility in regulations can be largely a response to mistrust: if an environmental organization or government agency distrusts the regulated entity to meet the spirit of the regulation as well as the letter, it will write the letter of the regulation so rigidly that gaming is impossible. Manufacturers have, in some circumstances, been extremely creative in their development of ways to avoid the intent of regulations, while still technically compliant.

Perhaps the worst example of circumvention is an auto manufacturer that developed a switch or valve whose purpose was to operate only during the fuel economy test and to change some of the settings in the engine to produce a higher fuel economy rating using the test protocol. However, the valve would never be used in actual driving operation, so it clearly was intended to do nothing more than produce a bogus result. More recently, another manufacturer attempted a similar trick for refrigerators.

Another example of gaming with compliance is the first generation of energy performance standards promulgated by the California Energy Commission (CEC). In the mid-1980s, home-building consultants developed a technique that assumes the reference house—the house whose energy efficiency the actual house had to meet or exceed—was less efficient than it actually was. The trick was to assume that the reference house was built without a structural concrete floor (most California houses are built on slabs on grade and a concrete floor is necessary for the physical stability of the house). Yet the slab floor not only acts to hold up the house, but also to provide an improvement in energy efficiency, which was ignored in the calculation of the energy efficiency of the reference house. This technique reduced the energy efficiency of the reference house by about 20 percent, similarly diluting the effectiveness of the standard. The CEC later eliminated this loophole with the somewhat shamefaced support of the building industry.

The experience with closing the loophole with the California building energy-performance standards illustrates a way out of this bad feedback loop, but one that is seldom used. If the problem is that flexible compliance mechanisms can create loopholes, one solution may be to make regulations inflexible. But another may be to develop regular review processes that identify known loopholes and close them. This approach requires more staff commitment by the regulatory agency and by the public-interest watchdogs, however.

It is worth noting that the same problem can also happen if the writers of the regulation do not understand the needs of business. Regulations, unfortunately, can be inflexible as a result of poor drafting just as easily as they can be inflexible in order to prevent gaming.

The Costs Imposed by Sudden Changes in Regulation

Another major concern of industry is that uncertainties about future regulations can produce additional economic cost. If, for example, a chemical company plans to design a new production facility, but is uncertain about what level of air-pollution and water-pollution control will be required, then it has no idea how to design the plan or how much money to invest in it. In many cases a known answer, even if it's less desirable, is better than continued uncertainty.

In a worst-case scenario, the plant designer could plan for a modest level of pollution control, discover that the standards increased during the construction period of the plant, and then could have to face an expensive retrofit project of the "new" plant and perhaps suffer down time while the retrofit takes place. In another scenario, the regulations could get tougher ten years into the forty-year design life of the facility and render it obsolete.

Industry's apparent desire for certainty, which many companies have repeatedly expressed explicitly, does not, unfortunately, seem to be reflected in its strategic behavior. When the 104th

Congress, a much more conservative and anti-environmental-ist crowd than the previous Congress, came into power, House Majority Leader Newt Gingrich proposed massive rollbacks in environmental laws and regulations. Industry by and large supported these proposals.

Such a cycle repeated itself within many industries during the George W. Bush administration. In the case of this administration, many of the executive branch's proposed environmental rollbacks were a direct response to industry requests.

Yet the type of advocacy described above is directly opposed to industry's strategic interest in regulatory certainty. First, an industrial company that already planned to comply with the regulations would be faced with uncertainty if its colleagues or even its own corporation suddenly were to lobby for regulatory relaxation. Should the engineers assume that these lobbying efforts would be successful and plan for weaker standards? Or should they assume the status quo? Or just delay the investment until the dust settled?

From a broader strategic point of view, the environmental problems that the regulations are designed to address aren't going to go away if ignored. It would be prudent for industry to assume that a problem unaddressed in 1994 would be addressed again when a more pro-environmental administration came into power or when states took matters into their own hands in the face of federal inaction. It is similarly prudent to assume that a rollback today will be reversed when someone with greater environmental sensitivity is in the White House. Wouldn't the cost of this uncertainty outweigh the economic benefits, even assuming there were any, of the regulatory rollback? American industry has failed to present a consistent view on these issues.

The Need for Better Communication

Many of these problems of distrust are the result of inadequate or nonexistent communication between industries and their trade organizations and the environmentalists. Distrust obviously is generated most easily when one can't even hear the other side's point of view.

My experience is that efforts to find win-win solutions through negotiations are successful more often than not. In many ways environmentalists can help solve industry's problems in a socially beneficial way if both sides recognize this as a goal. For example, the auto industry has argued for over twenty years that increased fuel economy in cars is not practical for it because the industry sees no consumer demand for higher fuel economy.

Superficially at least, the industry perspective is accurate. If a car manufacturer offered a vehicle that saved 20 percent of the gasoline used and cost more by an amount that would be paid back in three years of driving, the car probably would not sell. Yet this problem can be framed as a marketing challenge for manufacturers, if they wish to do so. The problem could be framed as, "we would be happy to produce more fuel-efficient cars if you can assure us that there will be market for them after they were produced."

If such a marketing challenge is accepted, possibilities are available for collaborations that can solve the problem. This was done in the case of clothes washers, where a nationally consistent program specification for efficient clothes washers made it profitable for industry to introduce a new generation of products that were more profitable, offered better consumer service, and incidentally saved two-thirds of energy use and more than half of water use.

Better communication could lead to commitments by industry as well as environmentalists to negotiate in good faith over future policy directions, with the goal of establishing agreements with ten or even twenty year time frames that give industry a clear pathway to travel where it will receive support from government and public interest organizations.

And clearly both sides need to be sensitive to the issue of how to create perceived trust, and take whatever steps are necessary to assure that agreements are kept and that the organization has enough institutional memory to know what it is it agreed to do.

Two examples in the next chapter explore the themes of weak communication channels and mistrust. The examples illustrate how these problems have created political controversy over global climate change in America that is unique in the developed world, and how the process of nongovernmental regulation generates unnecessary friction between environmentalists and business. I also suggest some possible explanations for these problems.

10 | What Truly Motivates Anti-Environmentalists

In this chapter I offer some thoughts as to what motivates anti-environmental forces—and what the source of their political strength is.

Previous chapters outlined a world-view that is based on myths of what environmentalists truly stand for and how economic fundamentalism could be seen as a tool to defend a business interest that believed in these myths.

In this chapter I highlight alliances of convenience between economic incumbents who benefit from the maintenance of the status quo, even when doing so fails to deliver economic growth. I also discuss ideologically motivated people who derive pleasure from the belief that they understand how the system works better than others.

Recall Chapter 5, in which I introduced the strength of such an alliance of convenience in which a coalition of large electricity consumers in California who objected to high electricity prices and of free-market experimenters led to the state's disastrous attempt to restructure the electric sector.

This chapter explores another example of such an alliance of convenience—between defenders of conventional wisdom in the economics profession and companies who benefit from the continued use of fossil fuels. The argument is about the appropriate response to global climate change.

A Story of Energy Efficiency and Global Warming

The observation that energy efficiency promotes economic growth has profound consequences for the economics of controlling global climate change.

The emission of several pollutants that trap heat in the earth's atmosphere causes climate change. A single greenhouse pollutant—carbon dioxide—causes over 80 percent of the problem. Carbon dioxide is produced whenever fossil fuels are burned, so to control climate change requires that we reduce energy use, or else capture the emissions and store them underground.

Scientists have become increasingly concerned about the consequences of climate change and the fact that the problem becomes increasingly severe the longer we procrastinate taking action. The United Nations led an international negotiation to slow climate change that in 1997 resulted in the development of the Kyoto Protocol, an international agreement that requires the developed nations of the world to cut their greenhouse pollution emissions by about 5 percent in 2010 compared to 1990 levels. This small reduction, which doesn't even apply to the whole world, is a critical first step toward a longer-term, more ambitious goal for emissions reductions that will be negotiated in the future.

While most of the nations affected by the Kyoto Protocol accepted the agreement—it went into effect legally in February 2005—the United State has not accepted it.

Political opponents of the Kyoto Protocol have tried their best to create the perception of scientific doubt about the seriousness of the problems of climate change. Over the years, the arguments of these so-called skeptics have become farther and farther out of the scientific mainstream: increasing evidence validates the seriousness of the problem of anthropogenic climate change.

I omit in this book the scientific issue of how serious such a threat is, and instead look at the more limited issue: "how would compliance with the Protocol affect the economy?"

This limited approach makes sense because if it is true that to mitigate the threat benefits the economy and promotes growth, then it doesn't matter how serious the threat of climate change is: we should try to avert it in any event.

How Kyoto Would Effect Our Economy

The main reason—perhaps the only reason—that the United States has not agreed to Kyoto is that American political and business leaders are worried that it might have an adverse effect on the economy. Numerous criticisms have been directed at the Kyoto Protocol based on economic models that predict it will reduce economic prosperity in the United States. In fact some economic models give substantially different results (which should give pause to those who believe in them, but doesn't) yet nearly all of the purely economic models project that larger or smaller sacrifices are necessary to bring the United States into compliance with the Kyoto Protocol.

However, none of these purely economic analyses looks explicitly at the technical potential for energy efficiency. As far as these models are concerned, there is no difference between a scenario in which there are no technical opportunities to improve energy efficiency that offer a reasonable return on investment and a scenario in which economically preferable energy-efficiency options could cut energy use by 50 percent at a profit.

Yet obviously it makes a difference whether technical opportunities to reduce emissions and save money are available. The U.S. government and the academic energy modeling community have largely ignored analyses of energy efficiency. This is a shame because the majority of these analyses illustrate what *America's Energy Choices* shows: that the actual cost of compliance with the Kyoto Protocol is less than zero[1] for the United States. Compliance with Kyoto would actually reduce overall energy services costs and increase growth and jobs.

In addition to *America's Energy Choices*, a second study worth a mention is the U.S. Department of Energy's (DOE)

Energy Innovations study, which the national laboratories performed. This study showed more modest results, but this modesty largely was a consequence of the studies having made extremely conservative policy judgments about what levels of efficiency improvements were practical and likely to be adopted.[2]

Notwithstanding its conservatisms, the study predicted that more than half of the savings necessary to meet America's Kyoto Protocol goal could be achieved using identified cost-effective efficiency measures that could be acquired by known, reliable policy mechanisms.

An additional study builds on the *Energy Innovations* study in an interesting way. It couples the use of broad-brush economic incentives with the specific efficiency policies in the *Energy Innovations* study. The study examines the effect of shifting taxes from the employer portion of the Social Security tax, which would be eliminated, to a tax on greenhouse gas emissions, which would be established. The net impact on government revenue would be zero. This is a tax shift away from the taxation of things that we want more of (taxes on employment) and toward the taxation of things that we want less of (taxes on pollution).

The study reviewed the impact on industrial competitiveness of this combined tax shift, industry by industry. It revealed that for over 99 percent of the economy, U.S. businesses became more economically competitive with foreign businesses under the tax shift scenario than they were in the base case. Combined with the large economic savings from the efficiency policies already in the DOE study, this showed that well over 99 percent of the U.S. economy would be definitively better off under a scenario of compliance with the Kyoto Protocol. The benefits to this large majority of firms and individuals were so large that it would be possible to easily "buy out" the small fraction of losers and assure their undiminished welfare as well.

Note that this happy result—that 99 percent of the economy would be better off if the United States implemented the Kyoto Protocol, and that the other 1 percent could be made whole if we shared everyone else's profits—is based on several very conservative assumptions, namely that:

- the maximum levels of efficiency that could be achieved by policy were those in the DOE-funded study, rather than the environmental organizations' studies; and more importantly,

- none of the economic growth-inducing effect in the strong form of the growth proposition is counted. In other words, no new carbon-saving technologies are introduced, the cost of existing ones does not decline, none of the efficiency measures has nonenergy benefits, and no enhancements of competitive forces are induced; and

- the reduction in energy demand does not lead to any reductions in energy price.

In summary, a long list of detailed and careful studies by recognized experts in energy efficiency have concluded that relatively ambitious energy-efficiency goals—such as compliance with the Kyoto Protocol—can be achieved at a net profit to society, a profit that does not even include the value of the other environmental benefits of reduced energy use.

The level of net benefit is high—$2.3 trillion in *America's Energy Choices* ($5 trillion in energy savings less $2.7 trillion in efficiency investments), not even counting the indirect benefits listed in bullets above. These indirect benefits are likely to be even larger than the direct benefits of $5 trillion in energy savings, and are augmented by the fact that the $2.7 trillion in additional costs of efficiency are likely to be smaller than that—or even zero—as discussed in Chapter 3.

However, such a conclusion is not what this chapter is about. This chapter addresses the questions of why such information is unknown to U.S. policymakers at the federal level, and why it is rejected summarily in so many policy discussions.[3]

The Influence of Economic and Ideological Incumbency

If the most likely results of compliance with the Kyoto Protocol are enhanced economic growth and peripheral benefits to national security (lower oil dependence and lower world oil prices, for example) and environmental quality, why should anyone be opposed?

If so much evidence is available that compliance with the Kyoto Protocol would help the economy, why is it so controversial? Note that a resolution that encouraged the United States not to sign the Protocol unless developing countries were also subject to binding commitments to limit their emissions—and that expressed concern over "serious harm to the economy of the United States"—passed the Senate 95 to zero in 1997.

Here are three potential answers to the above question:

1. It's heresy. The studies on how and why compliance to Kyoto could be profitable have been virtually ignored by the economics establishment, particularly energy modelers. During the past ten years, in particular, the U.S. Energy Information Administration (EIA) has performed a number of studies looking at the potential effects of compliance to the Kyoto Protocol or of meeting other environmental goals. Both in its modeling efforts and in its list of references, EIA essentially pretends that these studies are completely nonexistent. The EIA neither attempts to refute them, nor acknowledges them at all.

Such action (or inaction) described above may be due to the attractiveness of the economic fundamentalist model among economists and policymakers. The end-use-based approach is ignored not because it is merely suspected to be incorrect, but because it is heresy. If economics is interpreted as a fundamentalist religion, then alternative approaches are not merely wrong, they are heresy. Heresy need not be refuted: it should be ignored and suppressed.

With a significant portion of the research and policy community ignoring or attempting to suppress reports suggesting that it would be easy and profitable to meet climate change protection goals, this position is not often heard or well understood at the policy making levels, such as in the U.S. Congress or at the senior levels at the Department of Energy. The evidence that senior policy-makers hear consists almost exclusively of the predictions of those models that say overall economic loss would occur from compliance with the Kyoto Protocol. Thus the fundamentally bogus argument of an environment/economy tradeoff is still dominant.

2. It favors the competition. A second argument anti-environmentalists and business organizations employ against the Kyoto Protocol is that, whether or not meeting climate change goals is good for the economy overall, the effect on export-dependent or import-vulnerable industries or on those that compete with foreign companies in countries not subject to Kyoto limits (such as China) could be adverse. Such companies could be assumed to be primarily economic incumbents.

The above argument, which is basically plausible but wrong nonetheless, goes as follows: if the United States complies with the Kyoto Protocol, one portion of the compliance strategy will involve the equivalent of a carbon tax—some policy that raises the price of climate emissions and thus the price of energy.

Industries in the United States that compete in international markets will be burdened with higher energy costs compared to their competitors, particularly competitors in countries like China that are free from greenhouse gas emissions limitations. Because relatively small changes in price can affect competitiveness, the concern is that unless all countries accept climate change prevention goals, American industry will be put at a disadvantage.[4]

Yet while this argument is correct superficially, it ignores the fact that the U.S. government freely chooses the specific policies to comply with climate stabilization initiatives. If we are concerned about the competitiveness of U.S. industry, several options are available to assure that climate protection measures are competitively neutral or even favorable for American industries.

The first such option is to use some of the proceeds of the emissions charges to help industry become less climate polluting through investment in energy efficiency. Numerous success stories of utility-based programs that have helped industries find and invest in energy-efficiency measures with excellent returns on investment have been mentioned in Part 1 and its references. If part of the American climate protection policy were to provide such assistance both in terms of technical consulting services and in terms of actual financial incentives to make investments, the industries that are at risk (and other industries as well) could actually find their energy bills reduced compared to the base case.

Secondly, a tax shift from the employer portion of the Social Security tax to a greenhouse gas-emissions tax would by itself improve the competitiveness of almost all segments of industry while providing the direct economic incentive to reduce greenhouse gas emissions[5] And the limited number of companies that were still losers could be made whole by transfer payments.

So overall the argument that to agree to the Kyoto Protocol would harm U.S. industrial competitiveness just doesn't stand up to scrutiny. Clearly if a carbon tax that would affect about 7 percent of gross domestic product (GDP), namely energy costs, can hurt international competitiveness, a labor tax (Social Security and Medicare) that affects some 65 percent of GDP, namely labor costs, is even worse, so its reduction would help the economy be more competitive and grow faster.

The current American policy to tax businesses based on their labor costs is a significant problem for industrial competitiveness. It is surprising that the same business community that is afraid of a small tax on carbon or energy use would not be stridently opposed to the current system of Social Security and Medicare taxes, which imposes a higher rate of taxation (almost 10 percent of salaries) on a much larger cost center.

The inconsistent response just mentioned is all the more surprising because the Social Security and Medicare tax burden rises continually, and because the issue of Social Security was put on the table politically in 2005. If political conservatives can raise the issue of how Social Security benefits should be changed, surely business people could advance the idea that Social Security taxation of businesses should be changed.

3. It's bad for the oil and coal industries. The third argument against U.S. participation in the Kyoto Protocol is that specific industries whose products are directly responsible for climate pollution, in particular coal and oil, would be hurt. Clearly these industries have spent substantial amounts of money and effort to sponsor the work of researchers who try to undermine, at least in the public eye, the scientific basis for worrying about global climate change, and also fund public information and lobbying efforts that highlight their position that meeting climate change goals would be prohibitively expensive.

It's understandable that industries like coal or oil or electric utilities (but see Chapter 11 for why utilities might be unconcerned) would attempt to protect their perceived, and possibly real, self-interest in such a way. What is surprising is that the rest of the business community as a whole would buy into this argument, which is fundamentally anti-competitive.

Why shouldn't the natural gas industry, for example, see restrictions on greenhouse gas emissions as something to its benefit—particularly when one of the obvious ways to reduced greenhouse gas emissions is to switch power plants from coal to gas? Why wouldn't the industry recognize it as a marketing opportunity to replace inefficient uses of electricity, such as space and water heating, with lower-emissions applications of natural gas?

And why would the rest of the business community, which stands to make far more money from efficiency and renewable energy production or from additional domestic spending generated by energy savings fail to recognize their self-interest in supporting climate protection policies that generate investments in efficiency?

Why wouldn't that portion of industry that works internationally—to sell into or compete in markets of countries that comply with Kyoto—choose to have a globally consistent business strategy, one that can avoid having to tweak production one way for U.S. markets and another way for everyone else?

The fact that such problems occur suggests that the business community has bought into the arguments of economic incumbents who could truly be harmed by controlling climate pollution; in part because these arguments can be framed by the economic incumbents in a more politically acceptable form of support for free markets rather than protection of narrow entrenched interests.

The concept of self-proclaimed pro-business positions being in fact defenses of economic incumbency is not limited to the environment. Dominant corporations or trade associations that lobby successfully to maintain their dominant positions and suppress competition are a common problem throughout the economy.[6]

It is easy to see why such lobbying may be successful when it's the government that is lobbied, but it is another thing entirely to explain why corporations or trade associations have been successful with other businesses.

Who Writes the Regulations

One of the ways that dominant companies can protect their positions is to write regulations in ways that serve their needs better than the needs of start-up competitors. This is possibly one reason why regulation is at the heart of most environmental debates.

Economic incumbents often hide behind the economic fundamentalist reactions against regulation in their public statements. Almost all testimony that I have reviewed from companies that oppose particular regulations has railed against the concept of regulation entirely, rather than focusing on the actual levels proposed or the structure of the rule.

Yet the real disagreement is far less about the question of

whether to set regulations, and far more about the question of who sets the regulation.

The Issue of Control

The California Energy Commission has attempted to regulate the efficiency of air conditioners and water heaters over the years in ways that their respective trade associations disapproved. In a recent case on air-conditioners, the criterion that the Commission wanted to regulate was one that was already being measured in the current process of testing every single air-conditioner product in the first place. A similar story held true years earlier for water heaters: the criteria that the Commission wanted to regulate were ones that were routinely derived from tests being performed at every water heater manufacturer. Yet both trade associations fought the Commission vigorously to restrict the use of a different criterion than the one the federal government required.

A key point of the controversy was whether the Energy Commission could collect data directly from manufacturers on the performance of their products on both criteria. This dispute actually went to litigation.

The issue could not have been expense or burden because, in the case of air-conditioners, the association has agreed subsequently to compile the data in question and establish a certification procedure to assure its accuracy for only a modest expense—almost certainly less than the cost to pursue the litigation. The only plausible explanation is that this is an issue of control.

Government-mandated standards for such products as air conditioning and heating equipment have also been the subject of severe controversy and frequent litigation. Yet the professional society ASHRAE (the American Society of Heating, Refrigerating, and Air Conditioning Engineers) has set voluntary standards for these products since 1975. States frequently adopt these voluntary standards and enforce them as building codes, and they are widely considered as de facto mandatory minimum standards for the whole United States, despite their lack of legal effect until recently. Yet the same companies and trade associations that supported the ASHRAE process and praised the standards that resulted from it have been almost unanimous in their vigorous opposition to government standards. Clearly the issue is one of control: in one case in the 1980s an air conditioner manufacturer lobbied the California Legislature in support of a bill that would have restricted the Energy Commission from enforcing an equipment efficiency standard that differed from the ASHRAE standard.

A skeptic might argue that another important difference is stringency. In many cases, the government standards for products for which there were also ASHRAE standards have been more stringent. But unless the evidence demonstrates that stringency is in opposition to the economic interests of the companies involved (and it should make no difference to the trade associations), this argument falls flat.

Based on the last Section's discussion of risk aversion, loss aversion, status quo bias, peer pressure, and the problem of agency, it is apparent that a trade association–dominated process to set standards would produce much less aggressive results—those that cause less change in the status quo—than a government-sponsored process that looked in more detail into the cost and benefits of different levels of standard. But these same biases suggest that it is at best unclear which option is in industry's economic self-interest.

An odd illustration of the control-versus-self-interest issue comes from the early 1980s, before the 1985 discovery of the ozone hole that ended the political debate over whether to do something serious to control ozone depletion. Some of the strongest lobbying voices against restrictions on ozone-depleting pollution were the chemical manufacturers that produced chlorofluorocarbons (CFCs), the largest source of ozone depletion.

One of my colleagues had a chance to talk with an executive from one of these chemical companies. He discovered that the company's profits on its existing line of CFC products were declining, while a new line of proprietary substitutes was expected to have larger markets and higher profitability. Clearly the company would have made more money if they had just allowed a phase-out!

And in retrospect, we find that chemical companies make money as they sell the new chemical substitutes for CFCs —although not as much as they would have had they acted sooner, as the added delay and more stringent regulation that followed resulted in much greater incentives for investment by competitors and customers, who successfully developed many substitutes that did not require proprietary chemicals.

When asked why they had maintained such a seemingly self-defeating stance, the company replied that regulation of CFCs might make regulation of greenhouse gases (a more troubling proposition) more likely —without ever recognizing that both cases presented opportunities for new markets and profits.

So much for the myth of the greedy corporation.

The real problem in the case of CFCs was not profits, but rather control.

So we see from the examples presented here that the issue is not one of regulations versus free market. At most, it is a policy issue of the role of government in deciding what the regulation should be rather than relying on private-sector processes.

To frame the question not as one of markets versus regulation—and instead as the issue of who is best positioned to decide what the regulation should be—gives a much different and potentially less contentious answer. Sometimes the cost and bureaucracy of government standard setting is not justified by the public benefit. And sometimes the degree of public participation and third-party judgment provided by a government standard-setting process is essential to the achievement of a result that advances technology and promotes competition.

Government Versus Private-Sector Regulation

It seems evident that when the following set of conditions is encountered, government regulations would be better than private-sector regulations:

- the regulations have a strong effect on the public interest
- the regulations affect the environment to a meaningful extent
- the regulations have a major influence on the well-being of consumers
- the regulations influence the level of competition within an industry—and particularly the ability of new companies to come into markets
- a particular choice of regulation affects some regions of the country or world differently than others, or some income classes differently than others.

One of the key differences between private-sector regulation and government regulation is the process that is used to address technical and policy issues. For private-sector regulations, the key criterion is consensus. The American National Standards Institute (ANSI), which despite its name, is a private nonprofit organization that has no formal relationship to the U.S. government, defines procedural rules by which many standard-setting organizations operate. These rules focus solely on how to provide opportunities for public comment on draft standards and what criteria must be met to obtain a consensus in support of a standard from the committee members who are authorized to write it.

The concept of consensus, which sounds democratic, is less open than it seems. Committees must be balanced between industries, but the public may have little or no representation on a committee. So if the balance is decided to be between producers of a product, users of a product, and general interest, a committee of fifteen might have six producers, five users, and four members of the public. A vote of the eleven industry members for a standard could easily be defined as a consensus.

The role of the public may be compromised even further from this. The members of the public on a committee could include academics whose research is funded by the industry, and who may consult for the industry on the side.

An important weakness of consensus-based standards is that many important standard-setting processes and organizations have no substantive or objective criteria that standards must meet. So, for example, a standard on energy efficiency need not achieve any measure of energy savings, whether objective (e.g., a 20-percent savings) or subjective (e.g., best practices; or the maximum savings that are economically justified).

Worse, the review processes look at how well consensus is achieved, not whether the consensus is justified. Thus each time a draft regulation is written, it must be made available to the public for review. Anyone can file a comment and request a change. If the change is not accepted, the commenter must be contacted individually, and the drafters must try to resolve the comment. ANSI processes require that a dissatisfied commenter receives a chance to appeal a decision after a near-final standard is accepted.

However, such an appeal is granted or denied on procedural grounds only. A standard that declares "pi equals 3" or "the world is flat" cannot be challenged on the substance. So if a commenter's problem is that she knows the world is round, but she has had the opportunity to speak to the standards-setting committee on this point, the appeal will be denied.

The sort of process just described inherently favors economic incumbency because it is time-consuming and expensive to participate. The procedural requirement to try to resolve each and every commenter dissatisfied with a proposed decision, while it seems on the surface to be the process that is most fair, in fact rewards those interests that can afford to complain the loudest and longest, and file the most appeals. Small companies or public-interest organizations may lack the budget and staff resources to act similarly.

In contrast, most government regulations are set pursuant to laws that have defined objectives that the regulation must meet. Government regulators not only have to satisfy procedural requirements for open public processes, but also they must get the right answer according to the underlying law. If they fail to do so, the aggrieved party can seek remedy in the courts based on arguments on the merits. This process guarantees a more level playing field.

This is not to argue that governmental regulations are better than private-sector regulations. Economic incumbents or their allies often dominate governmental regulatory processes as well. Governmental processes are likely to be slower and costlier than private-sector processes. And in some cases the level of technical knowledge that private-sector players bring to the table makes it difficult for governmental officials or people outside the industry to add anything constructive.

The main point is not that one sort of regulation or regulatory process is better than another. Instead, the point is that regulations are essential to how a competitive free-market functions. The real debates about environmental regulation are not about whether to regulate but rather who writes the regulations and what the required criteria are.

Such debates should be guided by whether the structure of a given process is likely to give more weight to competitive forces (or whether it needs to), or whether economic incumbency is likely to bias the result.

Well-written regulations, whether developed by the private sector or by government, will achieve their goals.

11 | Well-Designed Environmental Policies

I maintain throughout this book that well-designed environmental policies can enhance economic growth by encouraging innovation, expanding competition, and overcoming
market failures. Additionally such policies can thwart emerging
tendencies toward central economic planning, tendencies that
also impede economic growth.

The observations I've supplied suggest that an aggressive
environmental agenda can be devised that should be able to
move forward with the active support of most businesses and
without strong ideological controversy on the political level. In
the case of energy policy, this agenda is well illustrated by the
path being pursued by California today.

Politically California's energy path has been developed by its
dominantly Democratic legislature working with its Republican
governor, with the active lobbying support of investor-owned
utilities, municipal utilities, environmental advocates, and significant business organizations. Substantively, the policy:

1. Prefers efficiency and renewable energy sources, and
   selects all resources in a least-cost fashion.
2. Empowers utilities to be part of this strategy by grant-
   ing them the authority to make resource selections (with
   regulatory oversight) and making it profitable for them to
   succeed.
3. Asks administrative agencies to promulgate and expand
   standards for efficient energy use of buildings and equipment and for greenhouse gas-emissions levels of cars.

4. Supports tax incentives for long-term advanced achievements in efficiency and solar power.

5. Supports smart growth development.

It is an intentional effort to realize the goal of deploying well-designed environmental policies to enhance economic growth and business development.

What is a well-designed environmental policy?

This question actually can be framed in two distinct ways. The first is easier for someone with a background in environmental policy to answer: "How can you design environmental policies to achieve their direct goals while you maximize the effectiveness of markets and avoid or overcome their failures?" This question involves the design of market-based environmental policies that are as competitive and growth inducing as possible.

The second approach to this question is: "How can you design policies whose primary goal is economic development and that uses environment as a means to this end?"

The answer to this question is difficult to articulate at present. Yet I believe that if we can make progress as a culture answering the first version of this question, we will in the process establish the type of communications that provide an answer to the second.

Current Environmental Policies

Most of the regulatory literature and the public policy debate focus on a supposed contrast between command-and-control regulation and market-based regulation.

This is not a constructive framing of the issue, however, as the reader can tell by looking at the choice of labels. What American desires to be commanded or controlled? In a market economy, who wouldn't prefer market-based approaches?

Command-and-Control

Command-and-control is a term employed almost exclusively by opponents of regulation, or of a specific regulation, and the word choice is probably intended to evoke in the listener the image of big-government, top-down planning (discussed previously).

Opponents of regulation will use the term command-and-control to refer to a regulation that requires a particular industry to install a particular technology for a particular use. Industry executives, they claim, will be frustrated whenever similar savings from the installation of a particularly expensive technology could have been achieved in an alternative way that isn't covered by a regulation and can be achieved at much lower cost.

Because the term is used in such a politically charged way, it is difficult to figure out what the users of this expression actually mean: they often use it in reference to regulations that are unrelated to such overly prescriptive rules.

Particularly few command-and-control regulations that actually prescribe technologies have been promulgated, and almost none of those in the last two decades.[1] Most real environmental regulations now set performance-based standards of one type or another. For example, air emissions might be required to be less than X grams of sulfate per kWh of electricity produced at a generating plant.

Market-Based Cap-and-Trade Regulations

The term market-based regulations has a similarly politicized ambiguity in its meaning. I discuss below some examples of what I believe should be called market-based regulations, but these examples may or may not be considered "market-based regulations" by those who use the term in their advocacy.

The types of market-based regulations their advocates discuss are generally limited to cap-and-trade systems, or their nearly identical twin, pollution taxes. A cap-and-trade system is one in which the environmental pollution from a particular corporation or facility, for example, emissions of NOX from a power plant, is capped at an assigned value. A company that fails to meet its cap must buy the rights to emit more pollution from a company that has done better than its cap. Companies are allowed to trade freely and have rights to emit their assigned (capped) level of emissions plus or minus any rights they buy or sell.

At first blush, the cap-and-trade system appears to be the most market-based system. It overcomes the pollution externality and makes no further interventions in the market.

However, after an examination at how the key assumptions for markets are violated, and how markets are subject to systematic failures that discourage innovation (see Part 1) it is evident that cap-and trade, however market-based it may be, will neither maximize economic welfare nor minimize costs of compliance. I will next discuss the strengths and weaknesses of cap-and-trade, and then proceed to explore other policies, including nonregulatory policies, that I assert are much more market-based (on the assumption that the word "markets" means the types of markets that truly exist, as opposed to the fundamentalist image of markets).

In theory cap-and-trade allows individual plants or corporations with particularly high costs of compliance simply to avoid having to comply and instead can pay for someone else to achieve the same overall savings. Cap-and-trade has proven successful in a number of instances where it has been tried, supporters claim. Perhaps most widely cited success is the sulfur dioxide emission limits that have been capped since the early 1990s in the United States. The actual price for an emissions permit is about a tenth of the originally predicted cost.

The example just cited is used to show the superiority of cap-and-trade over command-and-control approaches.

Reality, however, is not so simple. More than just two regulatory approaches exist, and command-and-control is not a true competitor to any other approach. The most common type of regulation is performance-based, regulations that set absolute limits on emissions or some other environmental damage but do not specify what technological steps a company or individual must take to meet the standards.

The differences between a performance-based standard and

a cap-and-trade standard are not differences of kind, but only differences of degree. The primary difference is that cap-and-trade systems allow a broader range of trading among facilities and companies, rather than between technologies in a given facility or between technologies that can be used for a given piece of equipment.

Such difference is often crucial to the health effects of the emissions. While some emissions, such as greenhouse gases (carbon dioxide), for example, cause damage globally, so reductions in emissions in one place are interchangeable with reductions in another; other emissions (such as small particles) cause local damage. A cap-and-trade system could easily produce large reductions in emissions in remote areas where they do much less health damage and no reductions, or even increases, in emissions in populated areas where these emissions lead to much greater damage.

Cap-and-trade systems are also subject to the most serious problems with test protocols. If you trade something, what are the regulations for how to establish the specifications for what is traded? An ideal cap-and-trade system would have to cover all sources of emissions for a particular pollutant. But often that's impossible because of the limitations (or costs) to test for compliance. If the tests do not permit government regulators or market overseers of the cap-and-trade market to measure precisely what is traded, errors will result. Most likely these errors will preferentially be in the direction that allows more pollution than was intended.

Cap-and-trade systems may also fail to produce the innovation-inducing benefit that underlies the strong form of the economic growth proposition. Under a cap-and-trade system, plants can avoid radical changes, such as process engineering-based changes, that would enhance both pollution prevention and economic efficiency, but rather could make the simple, straightforward, and less economically attractive (but easier) choices.[2]

The experience with energy efficiency and waste minimization/process engineering in general confirms this fear. To clarify, assume carbon dioxide (CO_2) is the pollutant being capped and traded. If CO_2 emissions are capped and traded, the market will adopt a price for carbon emission permits. This is a price that an industry can obtain from others as it reduces energy use and CO_2 emissions and sells the permits, or that it must pay for energy use and CO_2 emissions if everything stays the same.

So in terms of its effects on an individual corporation, carbon cap-and-trade is equivalent to an increase in energy price.

However, as documented throughout this book, energy price is a very weak motivator of investments in energy efficiency. Industries regularly overlook 30 percent, 50 percent, and higher rates of returns on investment in energy efficiency. A carbon tax that, in an extreme hypothetical case, raised energy prices by 100 percent would turn 40 percent ROI investments into 80-percent ROI investments. But it might not assure that many more of them are actually undertaken because market failures and human failures are the primary factors that impede them rather than the economics of the investment itself.

In this case, while cap-and-trade certainly would achieve the environmental goal (by its very structure, with the assumption that emissions could be measured and monitored properly), it would do so in an unnecessarily expensive way: a two-fold energy price increase rather than the encouragement of investments in increased energy efficiency and improved processes that are already cost effective. This observation explains why many proponents of carbon cap-and-trade systems in the United States, such as those in the McCain/Lieberman Bill of 2004, emphasize the importance of progressive energy policies as a supplement to cap-and-trade in order to achieve environmental goals in the most economically productive fashion.

The argument that cap-and-trade fails to promote innovation may be somewhat an artifact of the experience with cap-and-trade systems. Most of these systems have set relatively modest goals for emissions reduction compared to the more direct plant- or stack-based emissions limitation regulations.

Yet this could be corrected in the future: forthcoming cap-and-trade policies could be so stringent as to require the majority of emitters to look at more innovative and more economically productive investments in process improvement.

Yet the fact remains that if pollution prevention opportunities with returns as great as 160 percent annually (the Dow example) exist, the emissions price would have to be extraordinarily high to achieve the same result that more directed policies could achieve.

So while the cap-and-trade system for regulating sulfur dioxide may have produced much lower costs for permits than was expected, how can we say that a more directed system might not have produced even lower permit costs than that?

Future Environmental Policies

What should a more directed policy look like? As a general principle—one easier to articulate than to follow—environmental regulations and nonregulatory policies (notice how little of the discussion above of economists' and policymakers' ideas of market-based environmental policies has even addressed the issue of nonregulatory policies, despite all of their authors' focus on markets) should be based on the specific details of the market that policy attempts to influence.

Markets are not all the same in the real world. A policy that promotes preservation of wetlands will be dissimilar to a policy that encourages energy-efficiency retrofits of strip malls.

Market-based policies should:

1. Look at the dynamics of the targeted market—at what is observed to happen, not what is expected by simple theory.
2. Specifically target attempts to overcome or circumvent market barriers, market failures, human failures, and institutional failures, and to deliver the result that an ideal market would be expected to produce.
3. Recognize what markets already do well, and try to

increase reliance on those mechanisms.

4. Promote competition among different vendors, technologies, and design choices so the government avoids picking technology winners (or, worse, corporate winners), and instead allows competitive forces to work.

We have seen, for example, in the design of new buildings and appliances that the competitive forces that promote low purchase price are effective and strong. The market fails to promote options that reduce operating costs, but it is most effective in its ability to reduce construction or production costs. Therefore policies that constrain energy and allow price competition would seem to work well.

Performance-Based Incentives

A general position that has been surprisingly missing from the literature on environmental policy is the use of performance-based standards or incentives, rather than cost-based.

The term "cost-based" means straightforwardly that the incentive is based on the cost paid for the environmental investment. Perhaps the best-known example is the federal solar-energy tax credit of the 1970s in which a taxpayer was allowed to take a credit of 30 percent of the cost paid for a solar water heater. Cost-based incentives are easy to understand and administer; however, they undercut market forces and have always failed to achieve their objectives, as we have found whenever the incentive programs were followed with careful evaluations.

The term "performance-based standards" is better defined than terms like "market-based" or "command-and-control"; it means standards wherein compliance is determined based on performance—on how much energy is used or saved, or what the level of emissions from a process is. A performance-based incentive for low emission cars might be a payment of $500 for each car whose CO emissions are less than X grams per mile.

Performance-based standards or incentives are preferable to cost-based ones for obvious reasons. We don't care whether people spend money on pollution control or energy savings.

The only thing we care about is whether they achieve the goal of reduced emissions or energy efficiency. In fact, if they can meet the environmental goal and not spend money, as American refrigerator manufacturers did eight times in the last thirty years, that is even better.

Cost-based incentives for energy were first tried in the 1970s for utility-scale renewables, consumer-scale solar energy, and insulation. The federal government and some state governments offered incentives for investment in efficiency and renewables. These incentives were expressed as a percent of price. For example, in California in 1980, it was possible to obtain a tax credit for 60 percent of the money spent on a solar water heating system.

The experience with all of these incentives was unfortunate: for solar, they led to a boom/bust cycle featuring fly-by-night vendors and poorly performing equipment. A solar water heating equipment that previously cost about $3,000 could sell for $5,000 with a 60 percent tax credit and still cost the user less than it had originally. Investors and consumers were so tempted by the opportunity to save money on their taxes that they paid more attention to the procurement of the system, and in the case of corporate investors, to the accounting that could allow them to claim the largest possible cost, than they did to the performance.

A large number of new solar and wind systems were installed in response to the tax credits, but their performance was poor. Prices did not come down. After the incentive sunset, consumer experience was so bad that the industries nearly collapsed.

In the case of insulation, some $5 billion of federal support was allocated to individuals who invested in insulation. Follow-up studies failed to detect a statistically significant increase in energy savings, however, compared to what would have happened otherwise.

And tales of abuse abounded. One particular (and apparently true) story conveyed how a contractor whose customers could take advantage of a cost-based incentive for solar hot-water systems offered them the installation for an inflated price and then threw in a free trip to the Bahamas. The tax credit would be the specified percentage of the inflated price, so the consumer would appear to be better off to the extent that they paid more. And, or course, the contractor gets to keep most of the tax credit, as she sells the same number of installations at a greater profit. The policy objective was to promote the sale of more product, which achieves the goal to commercialize solar technology and reduce its costs. But the cost-based credit drove costs up—not down—and rewarded with the most profits those contractors who bid on how substantial a contract cost receipt they could provide rather than how well the system performed.

While the story above may be an urban legend, it is clear that similar tricks would be impossible to discover in an audit and would be most attractive to those people who thought to pursue them.

Subsequent Incentive Programs

After these policy mistakes were discovered and evaluated, subsequent generations of incentive programs were designed differently. The successful programs now are performance-based. In these systems, incentives are offered for products or buildings that meet a targeted level of energy performance.

Such incentives are fixed in dollar terms. Thus if it costs $4,500 to save 50 percent of energy in a new home, the builder might get (as an example taken from the 2005 Energy Policy Act) $2,000. Yet if another builder offered the same energy performance for $3,000, she would save the entire $1,000 difference and would still get the full tax credit. This provides all the right incentives for different vendors to compete on the basis of energy performance that delivers the desired result at the lowest cost.

Another aspect of understanding markets leads to the suggestion that performance-based incentives should not pay for the entire increase in incremental costs. Why not? First, this high level of incentive is not necessary. Analysis of the failures of markets in new construction illustrates that higher first cost is not the primary market failure. Much more important are the human failures of risk aversion and loss aversion by the builder, and the problem of getting the builder to focus on the issue of energy efficiency at all.

Lack of information on energy efficiency is another barrier. The design of the performance-based incentive overcomes this barrier. Because the incentive is based on energy performance, the builder must do a calculation of energy use just to demonstrate compliance. But this document also provides the consumer with energy-efficiency information that he can use with subsequent buyers, or that can be taken to the mortgage lender to qualify the buyer for a larger mortgage to cover the $2,500 of additional cost that isn't covered by the incentive.

Note that the performance-based system is resonant with conservative ideologies, as well as those of other factions. Conservatives often have opposed tax incentives for energy efficiency on the grounds that a tax credit reduces or eliminates the incentives for the market to minimize costs. This is because the government pays the incremental costs (in some cases fully; in other cases only in part). But under a performance-based incentive system, all of the incremental costs fall on the consumer or the producer, so the market receives its full incentive to minimize these costs while it meets the environmental goal.

A performance-based incentive that covers less than half of the incremental costs also places the right incentives on other market players. The consumer will know that he actually pays for the energy savings, so his reactions will reinforce the builder's motivations both to sell the house based on its performance (lower energy costs and higher comfort) and to reduce production costs, as refrigerator manufacturers did.

Performance-Based Regulation

Performance-based regulation works similarly. The structure of regulations can be analogous to performance-based building energy standards where builders or designers have the option either to meet the standards via the use of a fixed menu of energy-efficiency measures or to equal the energy performance of a building of the same size that did employ those measures but rather achieves this goal through a choice of different measures.

Performance-based methods have been remarkably popular where they are available and usable. For new homes, California and Florida have developed easy-to-use systems and, as a result, over 90 percent of new homes comply via the performance-based method.

Of course slightly less energy will be saved by the performance-based method than prescriptive compliance. In the case of prescriptive compliance, any feature of a building that accidentally is a little more efficient than the minimum required stays that way. In a performance-based system, this compliance margin is traded off against under-compliance in some other area.

Yet in practice, my observation is that an enhanced ability to enact a stronger standard in the first place trades off such a weakening. If those who comply recognize that a particularly innovative new measure is not essential, then they are more likely to accept it as part of the regulatory process. Thus in 2001, the California Energy Commission required the use of third-party tested leak-free duct systems as part of the standard without overly vigorous opposition from the building industry. This acceptance was solely due to the existence of not only one, but also several performance-based tradeoff options that allowed builders to use other methods to achieve compliance.

In general, the effort to update and improve the level of efficiency in the California building standards has attracted much less builder opposition, and much less political opposition, than have similar efforts in other states that fail to truly offer the performance-based compliance approach.

Some of the most successful regulations have been performance-based requirements that were established at advanced levels of technological challenge but that allowed industry ample time to comply—eight years, for example. These regulations may include a weaker, more immediate phase as well.

Long lead times allow industry to plan and experiment to find the most attractive technologies to meet the performance goal. They also allow markets to develop among component producers who may be able to help the regulated companies comply even more effectively than they could on their own.

However, the problem with long lead times is that not only do they provide corporations with the time to get their engineers and business managers to address the real problems, they also give their government relations people, lobbyists, and lawyers time to try to overturn the regulation. This is why the issue of trust is so significant in crafting regulations that encourage innovation and growth. In order to take the needed risks to innovate, companies must be able to make their investments to meet a standard with the confidence that no competitor (one that refrains from improving products and instead tries to turn back the mandate) will undercut them.

Well-Designed Refrigerator Regulations

The refrigerator story in Chapter 2 provides a good example of what well-designed regulations and incentives look like and how they have worked.

The California Energy Commission first promulgated refrigerator standards in 1976. At that time, refrigerators were available with a range of features and a variety of sizes, yet the larger and more feature-laden products used more energy. Some refrigerators required a manual defrost every few weeks; others required this procedure only for the freezer compartment and only one or two times a year. Some had a freezer at zero degrees Fahrenheit that could keep ice-cream solid and some only reached 15 degrees Fahrenheit. Some had top and bottom doors and some had side-by-side doors—and a few had a single door only. Refrigerator production was similar to automobile production, not only because of the energy variations but also because the bigger and more energy-intensive products were more profitable to build and to sell.

To respond to this variation in consumer choice, the Commission set a standard based on the energy use of the most efficient models that were already on the market. Each product class had its own standard based on the need for the added feature-laden products to consume more energy. The standard was written in terms of energy use per year. And the maximum level of energy use varied by size (e.g., units that contained larger freezers or larger units overall received a more generous limit).

Financial incentives for even higher levels of refrigerator efficiency, which were offered by utilities shortly after the standards took effect, were keyed to the standard levels, which preserved the neutrality with respect to consumer choices. Thus the target level to qualify for a rebate might be 20 percent lower energy use than the standard.

As standards were strengthened, the same structure was used. Different standards were established for different feature classes and new classes were added for such features as through-the-door ice service that was introduced over time. The standards continued to vary based on size.

Such a structure was based implicitly on recognition of where market forces worked and where they did not. Market forces encouraged manufacturers to reduce the cost of refrigerators, even before energy-efficiency policies were adopted. No reason existed to believe that market forces failed to work in terms of the range of features and sizes that were offered. Yet it was clear that market forces failed to induce the use of cost-effective efficiency technologies.

So the policies mentioned above were designed to encourage new technologies yet not to affect the other attributes of the product, and to harness competitive forces for cost reduction.

California's Construction Regulation

Similarly California's new construction standards were well designed because they offered a usable performance-based option. As in the case of refrigerators, markets did a good job of minimizing the costs of new buildings. Yet they ignored the opportunities for productivity enhancements that could be achieved in conjunction with energy-efficiency investments and designs.

The standards were drawn up to establish a cost-effective energy goal that all new buildings must meet. But the market continued to be harnessed to allow contractors to build at the least cost. If the prescribed efficiency measures were not the least costly way to save energy at a particular site, which it turned out they seldom were, the builder could do something else to save money and still meet the energy goal.

The California building energy standards worked in part because the state spent a significant effort to ensure energy use could be calculated accurately, based on the installation of products whose presence inspectors could verify. One of the weaknesses of performance-based regulation, and cap-and-trade systems in particular, is that regulators have to be able to measure the traded quantity. If errors or omissions in the accounting exist, these amount to loopholes in the regulation.

If a regulatory system is perceived as ineffective or unfair because of loopholes, the regulators will become disenchanted with it and may instead move to simpler and less flexible regulations that are easier to enforce. Advocates of cap-and-trade sometimes forget how important it is to assure the reliability of the measurement system.

Using Both Incentives and Regulations

Pro-growth environmental policy should embrace incentives as well as regulations. Chapter 6 introduces a number of market barriers, market failures, human failures, and institutional failures that cause individuals and corporations to under-invest in innovative technologies. The provision of information and performance-based economic incentives for higher levels of efficiency or other measurements of environmental performance can overcome these problems. For example, utilities in many parts of the United States offer consumer rebates for specific devices or whole buildings that exceed codes or standards by a specified percentage. Often these incentives are tiered, with higher payments for higher levels of achievement.

The programs mentioned above have proven successful in making markets work the way economic theory suggests they should. Their direct effect is to cause markets to select products that are, say, 20 percent more efficient than the relevant standard. But their indirect effect is even more appreciable: they lead manufacturers to innovate. In several programs, manufacturers have said to the utility, "If your current program offers a $100 rebate for a 20-percent savings, what if I can produce a new product that saves 40 percent? Could you offer $200 for 40 percent savings?"

The answer to such a question above is usually affirmative. Thus for one of the longest-established programs that encourages manufacturers to produce efficient clothes washers, the initial target efficiency in 1992 was a "modified energy factor" 1.26, but the current program for 2007 calls for three tiers of a "modified energy factor" of 1.8, 2.0, and 2.2, and with additional specifications for water efficiency. And the initiative for establishing the highest specification for water efficiency came from a manufacturer.

The Alignment of Private Profit with Societal Benefit

A key principle that enables these programs to succeed is to align private profit with societal benefit. In many cases, these goals currently are antagonistic, so it is not surprising that corporations confronted with these constraints make the wrong decisions for the environment.

A key example of perverse incentives is found in utility regulation. All states and most foreign jurisdictions regulate the rates utilities can charge for electricity and gas. The rates are calculated to allow the utility to recover the costs of their operation, and to earn a reasonable rate of return on capital.

Most of the utilities' cost is return on capital investment. So, for example, if the utility needs to recover $.04 per kWh for capital and $.03 for operating costs, the rate will be $.07 per kWh. But this system inadvertently creates a perverse incentive for uneconomic behavior.

Suppose your consumption goes up by 1 kWh. The utility incurs an additional cost of $.03 but it charges you $.07. The company makes a $.04 windfall profit.

Even if your additional kWh causes them to need to invest that $.04 eventually, the windfall profit remains. This occurs because as soon as the new power plant is built, its costs are added to the amount the utility is allowed to charge, and they recover the $.04 from future sales. The utility still makes windfall profits from additional sales and incurs parallel losses from reductions in consumption.

A utility regulated in such a way will lose money if it encourages its customers to save electricity at $.03 a kWh and avoids the need to purchase more at $.07. This is perverse behavior economically, but it is an inevitable result of traditional regulation of utilities.

More thoughtful regulatory schemes solve this problem with an alignment of societal benefit with private profit. They do this, in the example just discussed, by allowing the utility to collect only the incremental cost of $.03 for incremental sales.

This goal is accomplished by the requirement that the utility recover capital costs based on predicted sales rather than actual. So if sales go up by 1 kWh (compared to the prediction) as in the example, and the utility overcollects the $.04 (because its rate is still $.07), it will have to return that windfall to its customers (with interest) the next year via lowered rates. Conversely if sales are below forecast, the utility recovers the $.04 losses with raised rates next year.

Such a regulatory system makes the utility indifferent to sales volumes. It also makes more sense given the cost structure of a capital-intensive monopoly business—capital costs are recovered based on investments, in a way that doesn't depend on sales, and operating costs are recovered based on operations (sales). The revenue structure is aligned with the cost structure, and provides the utility and often its customers with greater stability along with other benefits.

And a revenue structure as described above can be combined with systems that reward utilities that run successful efficiency-incentive programs if they are permitted to pass along the cost of the programs and allowed to share a fraction of the net savings to the customer. Such systems are in operation in several large states, and the Edison Electric Institute and the American Gas Association, along with efficiency advocates, endorse them.

Utilities that are regulated in such a way have become recognized as world leaders for how they develop and implement efficiency programs. As noted in Figure 7 (Chapter 2), utility programs in California are credited with almost half of the total efficiency gain in the state.

Market Transformation

Another environmental policy principle that can support economic growth is called market transformation.

Market transformation programs encourage widespread sales of new technologies by targeting a technology that is expected to be cost-effective and offer financial incentives for the production or sale of products whose performance is based on the use of that technology. For example, the clothes washer program described above was an example of market transformation. The program established a specification for extra-high energy efficiency and water efficiency based on the use of a clothesbasket that rotates around a horizontal axis. In other words, the basket spins in a way that moves like in a dryer: the clothes are spun up and out of the water and then they drop back. Because the basket does not fill with water, less is needed.

Such a program was based on the energy savings, not the horizontal axis technology, which encouraged competition and innovation: several manufacturers turned out to meet the specification with washers that employ top-loading, vertical-axis designs.

Naïve economic theory suggests that market transformation is unnecessary. New technologies nearly always cost less as the experience to make them increases.[3] Theory shows that the market will price the technology at the eventual low price rather than the relatively high price of a technology without much production experience. This is called "forward pricing," and companies that want to get a foothold in the market (such as Japanese car manufacturers in the 1980s) have employed it on occasion.

Yet market failures generally prevent the occurrence of forward pricing. We observe that the prices charged for new technologies start off high and then decline with market success, in sharp contrast to the theoretical prediction. The temporary incentives in market transformation programs emulate the theoretical optimum-pricing levels and allow sales to grow until prices truly do decline.

Beyond these few simple principles, it appears that little attention has been devoted to the issue of how to craft properly designed environmental regulations. Even less attention has been paid to the question of how to design incentives for environmental protection. And particularly little attention has been focused on the issue of how to design environmental policies that actively and affirmatively promote economic development.

One evident conclusion this chapter offers is that an attempt to establish broad, general rules on how to make policies "market-based" won't succeed. Markets are not all alike. When they fail, they fail in different ways, and different policies may be needed to solve the problem.

12 | Where Do We Go from Here?

Throughout this book I have argued the proposition that well-designed environmental policies will enhance economic growth, and that they could do so even better if economic development were an explicit goal. This combination of economic growth and environmental protection policies would be easier to develop and implement if greater understanding and dialogue occurred between environmental advocates and pro-growth or business advocates.

Environmental Policy Promotes Economic Growth

How can environmental policy lead to economic growth? Here are several possible explanations:

- It can promote innovation as manufacturers or designers are required to learn how to work in newer, cleaner ways.

- It can overcome market barriers, market failures, institutional failures, and perhaps more importantly, human failures that systematically cause corporate managers to avoid risks, and are observed empirically to leave 30 and 50 percent and higher annual rates of return on the table, unexploited.

- It can create new competitive markets for achieving compliance with standards or targets in the future at ever-lower costs.

- It can open markets to new competitive entrants or technologies.

- It can resist the political/economic forces that suppress competition in order to benefit economic incumbents.

Environmental protection policy can promote economic growth through at least three different mechanisms:

1. It can encourage business to take advantage of static opportunities: ones that already exist—that do not involve new technologies—to improve environmental efficiency at a profit. In the case of energy efficiency, it can provide nonenergy benefits that accompany and may even be more valuable than the energy benefits themselves.

2. Environmental mandates and incentives can also help the economy to achieve the best possible result—the global optimum—rather than just muddling through with small incremental improvements. Imagine the effectiveness of the economy as a mountain range with a number of small hills as well as taller mountains. The best possible result would be to climb to the top of the tallest mountain. However, if the climber just muddles through and climbs upward continually rather than looking at a map, she is likely to climb only to the top of the nearest small hill without seeing the nearby larger mountain. This analogy is a good one because in some cases minor incremental changes cannot improve the current economic situation, but bigger changes can improve the economy dramatically. These situations occur whenever economic incumbents can use their political influence or market power to preserve the current market relationships and suppress new market entrants or new technologies or approaches. They can also occur accidentally.

3. Perhaps more importantly, environmental policy can create dynamic opportunities where the policies spur innovation that otherwise would not have occurred. Current business planning is based on the experiences of businesses that new technologies that offer economic savings don't sell. As a result, they fail to put any effort into product development for such technologies. If environmental

policy establishes an expectation that this will be different in the future, companies will look at even more advanced technologies, even if they know they will need policy assistance to sell them. Good examples of innovation are found in the product development histories of refrigerators and clothes washers wherein each time the industry was challenged to meet an ambitious standard and then provided incentives to go even beyond that, they responded with advances that went farther than anything that had been anticipated or even hoped for. As a result, new technologies appeared that performed even better than anyone had previously dreamed possible. We now see producers take the initiative to request that higher bars be established for incentives for products yet to be produced, but which they intend to produce if the incentive is established. This process strategically does an end-run around the problems of risk aversion, establishing an expectation that there will be special economic rewards for innovation.

Environmental policies not only encourage economic growth, but also can enhance economic freedom and democracy. Environmental policies can eliminate regulatory barriers to free choice embodied in both actual written regulations and informally established regulations. They can revise or replace those private-sector regulations that exclude the participation of those affected. The solutions may be more democratically written regulations that remain sponsored by the private sector, or may be regulations that government takes over.

Environmental policies can also help alleviate problems of poverty and reduce disparities in income or wealth. Worldwide we noted that countries that base their economies on resource extraction have the worst problems of inequality. Informally we observe that countries that are relatively resource efficient and that offer advanced environmental standards, such as Japan and Germany, have lower problems of poverty than other places with lax environmental laws. Also, such policies as energy efficiency help poorer people disproportionately. The poor spend the highest percentage of their income on energy: this group has the most to gain when energy demands are reduced.

Current Barriers to Environmental Policies

The public dialogue on environmental issues remains trapped in competing mythologies of anti-environmentalists and the environmental movement. Conservatives talk broadly about their support of environmental protection, but then fall back on ideological arguments that essentially say that any attempt to improve the environment will hurt the economy. Liberals support environmental protection in general, but then often throw in additional objectives, many of which are anti-corporate. These attitudes do not establish the basis for trust and civil discourse that might allow business and environmental advocates to craft more creative solutions.

The current policy debate has not led to what could be the most productive outcome: a discussion of the problem of systematic under-investment in innovation—not just in ways related to the environment but throughout the economy. It remains unclear whether the problem is even recognized as a drag on economic growth.

Virtually no discussion has occurred of how policy can stimulate or hold back innovation, or why the issue of innovation is important to growth. Outside of the energy area, almost no dialogue has taken place about market imperfections and what they mean for economic growth (or for environmental protection). Even in the energy field, the depth of the problems of the failures of markets—about market barriers, market failures, human failures, and institutional failures—has been underplayed.

Yet the existence and power of the failures of the market that I have described present a continuing problem: one that justifies policy intervention—not just once but on a recurring basis. These interventions are needed not only to solve the particular environmental problem at issue, but also to overcome barriers to growth.

Chapter 6 outlines how the problems of market imperfections are difficult to correct on a permanent basis. We may be able to overcome the problems with one targeted policy initiative in, say, 2006, but they will just recur in 2010 or whenever the innovative technology of 2006 becomes obsolete.

The strength and depth of these market defects explains why continually advancing environmental policy, as has been the case for refrigerators and for automobile emissions (in California) looks so successful in economic terms.

In this book I have focused exclusively on energy and environmental technologies, but the market's weaknesses, particularly the most troublesome ones, undoubtedly apply much more broadly. It is not only in the environmental area where 50-percent returns on investment are being overlooked. It is not only environmental innovations that are held back by risk aversion, loss aversion, status quo bias, and institutional failures.

The process required to solve the bigger issues is more elusive. However, in the case of environment, many clear and direct opportunities are within reach.

How to Transform the Political Debate

To see the environment as an economic development issue and the consequent attempt to correct these failures of the market could transform environmental policy and change the political debate on the environment. It could yield noticeably stronger (not to mention more sustainable) economic growth and a much more constructive—rather than antagonistic—discussion of environmental issues.

What are some of the elements of such a transformation?

Changing the Politics of Environmental Protection

Perhaps the first element of transformation would be to change the politics of environmental protection. The environment is usually seen as a left-wing issue: progressives support environmental protection and conservatives generally oppose it in any specific decision. Both the left and the right generally accept this connection. But both sides are wrong. I maintain that nothing is fundamentally left-wing about well-designed environmental protection policies, and nothing is right-wing in the opposition to such policies.

Conservatives might oppose badly designed environmental polices for different reasons than liberals might oppose them. Yet well designed polices should have universal support, simply because they work better.

What is the difference between a conservative approach to how to solve environmental problems compared to a progressive one? Conservatives often oppose regulation-based approaches and prefer incentives. But incentives cost significantly more than regulations, which is not the conservative philosophy.

For example, California spends roughly $25 million of state money annually to write and enforce its construction efficiency standards. The same level of market share for efficiency—nearly 100 percent—at the same level of efficiency probably could be achieved through exclusive reliance on economic incentives: performance-based incentives with the same energy goal as the standards. But to achieve this near-universal market share, the level of incentive would probably have to be about $2,500 per house, and maybe more. At current levels of construction, this amounts to more than $500 million annually.

If faced with the tradeoff between achieving an environmental goal through regulation at a cost to the government of $25 million and accomplishing the same thing through a voluntary, incentive-based approach that would cost the government $0.5 billion annually, it is unclear which is the more conservative choice.

Performance-based regulations apply proper incentives to encourage the market to work. Unlike cost-based policies that shelter consumers from the true costs of their decisions, performance-based regulations (or financial incentives) require consumers to pay fully for their choices.

So the simple hypothesis that conservatives would begin to

support environmental initiatives, but insist that they be based on incentives, while the left would argue for environmental initiatives based on government regulation, is incorrect.

Conservative writings that directly or indirectly link environmentalism with government-controlled central economic planning typically conclude with arguments against strong environmental policy. Environmental issues are perceived in their writing as a small battle in a large war on state-planned economies. Yet what would be the result if they realized that no linkage exists between environmentalism and Communism? What if environmentalists openly embraced the concept of the free-market system as the best overall way to organize the economy, and framed their arguments in terms of strengthening markets forces and promoting greater competition?[1]

If environmental policy is more than just a small battle in a larger war over economic systems, what would a conservative environmental policy be? How much would it differ from the policies I have recommended on energy and climate change?

Both left and right want policies that work—that produce the expected results (or better) and cost the least possible, both to the companies and households that are affected and to the government. But these are largely questions of fact, not of ideology, so there is reason to hope for more consensus-based solutions.

To divorce the environment from the left/right or Republican/Democrat context may be challenging for partisans of either side. Most environmentalists are personally progressive in their politics, and some may be happy to see the environment as either a wedge issue or an issue on which to seek alliances with supporters of other leftist issues. Many conservatives have so much invested in the myth that environmentalists are anti-growth or pro-central-planned economies that it may be difficult for them to embrace environmental policy as an economic development strategy regardless how convincing the evidence is.

Yet the benefits to the nation and the world from removing environment from the areas of controversy to an area of common interest are so large that it is worth some thought as to how we can do this.

Place More Emphasis on Strategic Planning

A second change could occur in the business community, where more emphasis on strategic planning would lead to better business management practices and completely different forms of political advocacy.

It's evident that business generally fails to invest in innovations that could provide rich returns on investment. How can businesses manage themselves differently to ensure these opportunities are sought and implemented on a regular basis?

Clearly the management strategy above would require more extensive use of engineering and scientific staff or consultants. But it would also require different reporting relationships or delegations of authority. What might these be? Businesses and business schools would do well to answer this question.

Businesses that lack the analytic resources to identify opportunities to be cleaner, more efficient, and more profitable likely also lack the resources to determine what government actions might actually help or hurt their businesses. The creation of a stronger strategic planning activity would help remedy this problem.

If businesses regularly oppose government actions that would increase their profitability, we have a long way to go in the reassessment of strategic goals and of strategic and situational allies.

I've found that business historically has looked at government as a dispenser of special favors or as a cop-like figure who might stop it from its preferred behavior. The view of government as a dispenser of favors in response to business lobbying is so well established that it is parodied in Mark Twain's first novel, *The Gilded Age* (published in 1871).

Business people pay almost no attention to the possibility that government actions, particularly regulatory actions, could contribute to a company's competitive strength and to its profitability while providing a level playing field to all competitors. And even less attention has been paid to the role that nonprofit organizations could play to balance competing interests and validate the public worth of a policy that profits a given company or industry.

Oddly one company whose executive I spoke to in the past year observed that a particular type of regulation could provide it an economic advantage, but "out of principle" the company would refrain from advocacy for that policy.

I believe this attitude is common. Yet perhaps better dialogue between business and environmental advocates would lead to the realization that the most ethical policy for a business is actually the reverse: if a policy is in the public interest, and it happens to benefit the company, all the more reason for the company to advocate for it. If the policy truly is in the public interest, that company ought to be able to find disinterested nongovernmental partners that can validate the competitive fairness of the policy and collaboratively advocate for it.

The company might even stand to reap public relations benefits for its support of a clean environment (or other societal benefit).

Alliances Among Business, Environmental Organizations, and Government

Another element of business strategic planning is to identify more pro-actively where environmental organizations or government can help companies achieve business objectives that are beyond the scope of an individual corporation's power. For example, when one washing machine manufacturer developed a more-efficient prototype, in contrast to the usual attitude of leaving the design on the engineer's shelf because a more expensive washer would be difficult to sell, it framed the problem as follows:

"We could produce a higher-value and more efficient product that makes us more money and makes our customers happier, but it won't sell in this market. How can we work with allies to make it marketable?"

By framing the question in such a way, the manufacturer

started to follow a new path. It consulted potential allies in utilities, nonprofits, and government and asked what the groups could do to encourage the sale of more efficient washers, should they become available. The manufacturer noted that a program designed to encourage a particular advanced technology could succeed because at least one company would be able to deliver that technology.

The above manufacturer found that the allies could establish and fund incentive programs and recognition efforts that seemed likely to work.

On a separate track, the company determined that a mandatory standard at a specific level of efficiency and a revised testing protocol that gave more credit for some previously unused efficiency options would also help improve profitability.

Other manufacturers reached more or less the same conclusion a little later. These more creative strategic plans led to the opportunity—one that several competing companies realize today—to make substantially more money from the same product by improving its performance so that it offered dramatically enhanced customer value. This last description sounds like something every company does; but what is different is that the strategy in this case required reliance on nonindustry allies to help create and transform markets. The mere decision to develop and produce the product was insufficient.

A New Role for Environmental Organizations

If business is able to make this transition, it requires that environmental organizations take on a new role. Competitive issues and even anti-trust law make it valuable for corporations that seek growth opportunities to have objective or financially disinterested third parties to share in the development of the new growth/environmental policies, or at least to vouch for these policies as being pro-competitive. In principle trade associations could play this role, but in practice the need for the third party occasionally to take sides with one company's proposal over another suggests that an outside (nonindustry) third party is best.

Some environmental organizations have done this for years, but it demands a much larger share of their workload if we are to solve the problems.

Part of the solution requires all stakeholders to look at the big picture, rather than to identify so-called bad guys to go after or alternately a few perceived good guys on the other side (from whatever side "we" reside) to work with. The process demonizing the opposition afflicts all sides. Yet to the extent that either side begins a respectful dialogue about common objectives, this problem can be greatly reduced. Business and environmentalists will not necessarily agree on everything, but to understand mutually the basis of the disagreements that remain helps stop the knee-jerk reactions and lays the groundwork to find additional common-interest areas.

Incentives and Regulation

A goal of this book is to encourage business leaders to think more strategically about how government environmental policies can shape a more profitable and successful business plan. These policies may include regulations.

Another goal is to encourage business schools to develop curricula and classes that provide the basis for the deeper analysis suggested above.

Finally, for the last three decades, environmental protection policy debates in America have focused most heavily on regulations. It is worthwhile to ask whether this is the best approach, and to answer the question based on facts rather than ideology.

Incentives have proven more effective than regulation in a few cases (recall how utilities can be rewarded for improving their customers' efficiency rather than being required to do so). In other cases, the tradeoff is more economic—either incentives or regulations can be effective, but regulations are significantly less expensive. Sometimes incentives do things that regulation is unable to: new technologies that are so advanced we are uncertain of their effectiveness obviously cannot be required, yet they can be promoted through incentives.

The best mix of standards, incentives, and other polices seldom has been discussed in practical terms—instead what little discussion there has been is generally so broad and philosophical that it fails to lead to any recommendations that can be implemented.

The Need for Cooperation

The first step toward solving almost any problem is to recognize that it exists. This book provides this recognition. One of my purposes is to challenge business leaders and supporters of economic growth to look in greater detail at their strategic interests and how they interact with the environment and in how they encourage innovation.

One thing that has become even more apparent as I wrote this book—and hopefully as readers complete it—is that environmental leaders and business or pro-business leaders fail to talk to each other. This is a serious problem because if one theme underlies all of the case studies and examples that have been reviewed, it is that overly simple solutions are unsuccessful. If we are to develop pro-growth and pro-environment policies that actually work, it will require active collaboration between business, government agencies, and environmental advocates, and will probably require pro-active discussions between progressives and conservatives.

What Is the Problem?

To identify a problem is easier than to identify a solution. And to establish policies that implement solutions is yet another step. It is doubtful whether any interest group can do it alone.

What is the nature of the problem?

From an economic growth perspective, the problem is under-investment in innovative technology. The relation of that technology to the environment is almost irrelevant from this perspective, except to the extent to which workers are more productive in a healthful environment.

However, the problem of nonperforming markets is deep and complex, and broadly focused policies that encourage corporate leaders to have greater risk-tolerance are difficult to describe. This difficulty is compounded by the fact that some level of risk aversion is rational. In fact, many of the failures of corporations to act in their own best interests that I have described throughout the book involve insufficient risk aversion to problems that may occur in the long run.

Another problem with the encouragement of innovation in general is that it appears remarkably similar to industrial policies in which governments try to "pick winners" from among technology options. Policies that pick winners often fail; indeed markets appear to work best in picking winners.

To use environmental protection as the driver of innovation overcomes these problems in that it selects technological problems that need to be solved on their intrinsic merits. For example, if over 30,000 Americans die annually from air pollution, then developing better emissions reduction technology is desirable even without consideration of its economic benefits due to the encouragement of innovation and competition. Establishing the technological specifications that the technologies need to meet is easier if our goal is to solve a specific problem such as air pollution: it can be done based on public health or toxicological criteria or other scientific criteria that make sense in their own right.

So if policy encourages efforts to solve this problem, it will move the economy in the optimal direction irrespective of whether the technologies that we think we want to encourage result in the best choices. The policy will be successful even if it fails to achieve an economic goal that encourages innovation because its direct benefits are important enough.

Yet this approach solves only part of the problem. We have observed that many current environmental policies have promoted economic growth more or less by accident. Few of these policies have been developed with economic growth as a direct or explicit objective. To change this would seem to require that we bring new expertise and experience to bear. This is one reason why the expanded dialogue between the various environmental stakeholders is so important.

To achieve such a dialogue in useful way requires all stakeholders to reframe the issue.

One obsolete frame is the question of whether environmental protection slows economic growth or is compatible with it. This frame suggests that pro-growth policies are of insignificance to environmentalists and that environment policies are outside the scope of a company's business plan. Framed in a better way, we ask how we can address environmental needs in way that strengthens markets—that makes them more effective at making the best selections.

Another obsolete frame is whether environmental policies are a creature of big government. Framing this way leads to the assumption that the real controversy is between advocates of market-based environmental policies, which are usually taken to mean cap-and trade, and advocates of command-and control, which is often used to mean anything that the user of this expression happens to oppose. One reason this framing is incorrect is that the biggest real controversy, in all the cases I have seen, is primarily over the stringency of controls rather than how they are achieved.

The last-mentioned frame is unconstructive because cap-and-trade is not the only market-based mechanism, and in many cases is not even a good market-based mechanism. Cap-and-trade only works if the specific market that is capped produces the results that theory predicts. For the case of energy, where the most controversial cap under debate is on greenhouse gas emissions, we know that the existing markets for emissions reductions (for energy savings) fail miserably, so to load more costs onto this structure will predictably also work poorly.

In such a case, to identify an effective market-based mechanism requires appreciably more work: we need to look at markets for motors and lights and air conditioners and cars, et cetera, and not just at markets for energy in general.

If the problems are due to systematic failures of current markets to deliver the expected results, then the failures transcend anything that an individual company does wrong, and its solution requires a great deal more than a few progressive companies that act differently. From the business side, while it may be great to do a signature green demonstration project with an environmental organization as partner, inadequate numbers of environmental organizations and staff are available to solve the problem this way, even if all of them wanted to work on such partnerships.

Less Ideology and More Results

Perhaps the broadest "big picture" suggestion I can offer is that all sides need to be less ideological and more results-oriented. Many environmental policies can be evaluated objectively with respect to their effects on the environmental problem at hand, and also with respect to the business prospects the corporations that will have to be involved in the solutions face. Direct economic stakeholders rarely perform these types of evaluations. Remarkably limited academic literature can be found on the subject of the measured effect of environmental policies on innovation, productivity, profitability, or growth.

The enhancement of innovation in the economy is a critical need both in the United States and to an even greater extent in developing countries. The reason that the global economy is in better shape now than it was twenty years ago, at least in the larger economies, is that productivity has increased. This increase is usually attributed to computer- or Internet-related efficiencies. Yet these seem to have run their course: little ongoing policy discussion occurs of how governments can encourage the expanded use of computers or the Internet to increase economic growth.

The U.S. economy, or that of any other country, could similarly reap the benefits of environmentally induced increases in competition or innovation, particularly if this were a consensus goal that could be implemented rapidly without the delays brought on by political controversy. Actions that were designed with both environmental and development goals in mind could enhance global prosperity while they also make the world a more healthful and better place in which to live.

There are good reasons for optimism about developing a pro-growth, pro-environment strategy. An increasing number of businesses, which range from small mom-and-pop producers to some of our largest multinational corporations, acknowledge that environmental protection is good business. Companies that are "greening" themselves recognize that good environmental attitudes help them to attract the best staff and give them an edge in customer awareness and satisfaction.

National environmental organizations have become more open to working in partnership with business, and a few distinguish themselves in their publicity by their pro-business attitudes.

However, such a movement toward cooperation and tactical alliance is incomplete. A business, no matter how green its intentions, will still be subject to all the failures of the market that were described in Chapter 6. Even if its environmental position enhances its profitability, it will still have to swim upstream: business performance will be less successful than if markets could work better or if their failures could be mitigated.

And such failures will frustrate environmentalists, as well. For every company they can work with for mutual benefit, another will exist that profits from the status quo, and will continue along its path of pollution for good self-interest reasons.

What is needed is a broader scale look at win-win solutions that embrace the whole market, or even the whole economy. Environmentalists and business can work together to correct the failures of markets and encourage innovation through jointly supported policies. These could be government policies—such as environmental standards, tax incentives or disincentives, changes in the regulation of utilities or other industries already subject to price regulation. They could be changes in industry-sponsored private-sector regulations, or changes to informal regulations.

The changes described above are far beyond the scope of what any business—even a large one—can do alone in a competitive economy. They often are even beyond the scope of what a whole industry can do through their trade association because they might appear self-serving or even anti-competitive to those outside that industry. Yet business/environmentalist or business/environmentalist/government alliances can accomplish these changes through cooperation.

The future success of the American economy—or of any country's economy—will depend increasingly on the substitution of innovation for resources as world business becomes more globalized and resources become increasingly scarce. To promote innovation in order to overcome resource scarcity is challenging to accomplish well with today's institutions. The failures of the market will continue to suppress innovation, and governments' inattention to resource usage will lead to continued resource depletion and its resultant impact on prices.

However, we know how to solve these problems. If business and environmentalists work together to do so, we can improve competitiveness and growth while we make the world a place we can leave to our grandchildren with a sense of pride.

Appendix

Myths and Realities in California's Experiment in Electricity Restructuring

This Appendix provides a close examination of the myths that economic fundamentalists have used to avoid their fair share of the blame for the failures of California's attempt to "create competitive markets" in electricity. It is important to examine these myths and their detailed refutation because they illustrate how naïve economic theory predicts outcomes that differ from what actually happen. The full story shows how real markets are more complicated than theoretical markets, particularly in regard to how they can fall apart under stress.

Myth 1: Huge Growth of Demand for Electricity

The statistics on electricity growth show that California electricity demand grew at a relatively constant annual rate of about 1 percent compounded throughout the 1990s. This is lower than the annual 2.2 percent growth rates experienced elsewhere in the United States for the 1990s (see Figure 8). In 2000 an increase of about 4.6 percent in electricity consumption occurred, about 2 percent higher than expected although not nearly enough to trigger a problem of the magnitude that was experienced.[1] Gowth in electricity demand indeed was a part of the problem, but some 85 percent of that growth in western electricity demand was outside California. (The eleven western states' electric transmission systems are highly inter-linked, so a shortage in one region or state becomes a shortage everywhere.) Planners in those states expected such growth, but California's energy planning process failed to consider it adequately. This growth soaked up the out-of-state surplus power upon which Californians had previously relied.

About the time of the California electricity crisis, a report was circulated that attributed something like 11 percent of electricity, nearly all of it recent growth, to computers and the Internet. But Lawrence Berkeley National Laboratory researchers, who observed that the calculation was based on a few highly erroneous assumptions, refuted the report definitively.

Most noteworthy of such assumptions was the idea that one could calculate the energy consumption of a computer if one looked at the power consumption on the nameplate and multiplied it by the number of hours the computer operated. In fact, the rated power consumption on most electronic equipment is no more than an electrical safety rating, and many such pieces of equipment never reach that rating under any conditions. And even more importantly such electronic equipment as computers operate at a variety of power levels, depending on how intensely the equipment is used. Average power use is typically less than 25 percent of the nameplate power rating, which ignores such energy-saving features as "sleep mode" that are incorporated into almost all new equipment.[2] Thus, when the report was recalibrated according to these facts, actual electricity use that could be attributed to the Internet and electronics, which included gross energy consumption only (that is, excluding energy saved in industry by the use of the Internet), was less than 2 percent.

Gross energy consumption is mentioned above because sometimes computers can substitute for other uses of energy and result in net decreases even though the computer itself uses energy. For example, when a file transfers over the Internet, it can replace physical transportation of a large paper document, which may never even be produced in the first place. The energy consumption of the computers must be balanced against the energy that would have been consumed to produce, print, ship, and eventually store the paper. Research by Japanese energy expert Haruki Tsuchiya showed that electronic publication uses less energy than comparable print publication.

So if high growth in California's demand was not part of the problem, what was?

Myth 2: Environmentalists and State Bureaucrats Blocked New Power Plant Construction

The second myth is that environmentalists or state bureaucrats were responsible for the dearth of new power plants. In fact, environmentalists who had opposed power plant construction in the 1970s had become supportive of construction by the 1990s. This is because while the 1970s power plant proposals were for high-pollution coal plants (to be built outside California) or for nuclear plants, proposed construction in the 1990s was for highly efficient and extremely clean combined-cycle natural gas–fired plants, or for renewable-fueled power plants. Environmentalists supported both options.

The reasons for environmentalist support for renewables are self-evident; the reason for environmentalist support for natural gas–fired power plants is that they are so much cleaner than anything else on the system that even if they displaced only a small fraction of the output from older power plants, the net effect would be a reduction in emissions.

The state regulatory problems with power plant siting and approvals were similarly overblown.

The California Energy Commission (CEC) had been established in 1974 in part to provide "one-stop shopping" to power plant developers to obtain state regulatory approval. The system, while more time consuming than in some other states, nevertheless provided assured approvals for needed projects in a timely way.

The CEC licensed twelve power plants in the early 1990s, and by 2000 nine were producing almost 1,000 megawatts of power. From 1991–95, environmental groups strongly supported CEC efforts and those of other state agencies to add another 1,400 megawatts of renewable energy and highly efficient gas-fired plants, but the Federal Energy Regulatory Commission (FERC) blocked the power purchase contracts that were pre-requisites to construction.[3]

The fact that the number of applications for permits dropped precipitously by the mid-1990s illustrates the problem was not due to environmentalist opposition or state regulation, that the problem was not that proposed power plants became stuck in the bureaucracy; the problem was that plants weren't being pro-posed at all.

What explains the fact that not a single power plant applica-tion reached the Commission from 1994–1997?

The answer is that investors in power plants had withdrawn their support. From the time at which the market perceived restructuring as a serious possibility (about 1992), until the time when the rules under which a power plant owner could sell electricity were finalized (in 1998), no one knew what to expect from a power plant in terms of revenues. Markets dis-like uncertainty—a project with a higher risk demands a much higher return on investment—so investors waited until the uncertainties were resolved.

Another sensible reason investors waited was because most market analyses projected surpluses of electricity and therefore low prices. Power plant developers could delay their projects until the uncertainties about the need for new power were resolved.

Unfortunately that took many years, during which time almost no new plants were proposed or built, and many planned plants were discontinued. Such an action is expected in a mar-ket economy.

Power plant construction was not the only victim of the uncertainty produced by the years of debate over the market structure of California's electricity system. Utilities had been investing heavily in customer energy efficiency since 1990 when the PUC changed regulatory practices to render such investments profitable to the utility whenever they were profitable to the customer.

Yet such investments raise rates (the price or tariff charged for electricity) slightly, and in some types of market structures, which included some under PUC consideration, such a rate disadvantage, even if small, is a competitive liability. As a result, utilities cut their budgets for efficiency in half after 1994. By 2000 this had cost the state some 2,000 megawatts compared to what could have been expected.

The widespread belief that power prices were low and declining due to the perceived power surplus compounded the risk factors. As noted above, the FERC denial of long-term contracts for 1,400 megawatts in 1995 was based on this (mis)perception. Neighboring states sold cheap surplus power to California, and apparently no one in either the power generation business or in state government realized that these surpluses were drying up.

Myth 3: Greedy Utilities Used Restructuring as a Plot

The reasons why this myth is untrue are complex; we must revisit the whole history that led to the restructuring debate. By the late 1970s, environmentalists insisted that investment in new power plants was unnecessary because it was less cost effective than investment in energy efficiency. (Subsequent evaluations have validated this belief.) After a few years, utilities, backed by state energy officials, accepted this argument as well and began to advocate energy-efficiency standards and support energy-efficiency incentives within their own customer relations operations.

That new power plants were stopped in the 1970s turned out to be remarkably favorable because the costs of the plants had escalated out of hand. The last proposal for a nuclear power plant in the state of California—the SunDesert plant near the Mexican border—had an estimated cost of $6 billion when it was finally cancelled. The state would have been saddled with an immense burden from high-cost power plants had construction gone ahead. And this problem would have been exacerbated by the fact that the power demand these plants would have been built to satisfy would not have materialized.

One of the other alternatives to conventional fossil-fired power plants was co-generation and renewable energy-based power plants. California implemented the national Public Utilities Regulatory Policy Act of 1978 in a more aggressive way than other states, a decision that spurred the creation of a renewables industry from a standing start in the state. The federal act encouraged state PUCs to require utilities to purchase power from independent power suppliers and pay them a fixed price for the output of these plants, to be decided by regulation, based on the costs utility would incur if it built a plant itself.

California enforced this legislation more aggressively by setting what turned out in hindsight to be an unrealistically high avoided cost for power—a cost that was based on the expected cost of plants that were almost as expensive as those that got cancelled. These prices were high enough to draw out a huge influx of proposals for new power plants.

This program was so successful that on the other hand California Energy Commission Chair Charles Imbrecht could boast with only slight hyperbole that under his administration, "California has produced more renewable energy for the electricity system, within each individual category, than the rest of the world combined." (In fact, a few categories existed where the rest of the world combined—but probably not the rest of the United States combined—produced more new renewable energy, but the main thrust of his statement was correct.)

Unfortunately for the state's economy, the unexpectedly high volume of applications led to a flood of new power plant construction that imposed a significant burden of fixed costs on the state's utilities.

When actual avoided costs went down compared to the prices offered under one of the so-called standard offers the CPUC devised, the "overhang" of expensive power caused electric rates to increase. By the late 1980s, rates were higher than they had been earlier, higher than they would have been otherwise, and higher than they were in other states. Of course, electricity supply was also higher.

Such a problem was greatly exacerbated as several noncancelled nuclear projects—long-delayed projects approved in the early 1970s—came into the utility rate base at the same time.

The resultant high electricity prices quickly caught the attention of California's large industrial electricity users who began to agitate both directly with the utilities themselves and indirectly through the political and regulatory system to get rate relief. Given the way utility regulation was structured, rate relief for some customers would obviously mean rate increases for others. Clearly to satisfy the big industrial customers was a political/economic problem.

The specific nature of the problem of high rates is that utilities were saddled with high fixed costs, which were due primarily to two factors:

- high-priced nuclear plants that were approved and built before energy-efficiency alternatives were considered

- high-priced qualifying facilities (QF) contracts, whose cost was based in large part on the expected high costs of coal and nuclear options

The fixed costs meant that rates would have to be higher than they would be if new gas-fired power plants were built because, since 1980, the cost of both the power plant and the natural gas to run it had declined.

Unlike for some industries in which high fixed costs mean losses to the company, for a regulated utility the rates are based on recovery of costs that are incurred prudently. Because the nuclear plants and the QF contracts all had advance regulatory approval, the law allowed the utilities to charge rates that would provide cost recovery, and the PUC set rates accordingly.

Proponents of restructuring believed that it would bring electric rates into alignment with the lower costs of new power plants, and that the reliance on competition among different companies to build them, rather than on utilities, would assure that the new, lower costs would get passed through to consumers.

Yet such proponents did not address the problem of high-priced nuclear plants and QFs: who would pay for these sunk costs? Because these costs would be impossible to recover in a restructured market and in competition with the new, cheaper gas-fired plants, they were also termed "stranded costs."

At this point in history the idea of deregulating electric markets, at least at the wholesale level, gained currency among both liberals and conservatives. The 1992 Energy Policy Act by the U.S. Congress encouraged the development of competitive wholesale markets for electricity.

Large-scale customers of electricity wanted to extend such so-called deregulation to allow them to shop competitively for electricity on the retail level. Under their proposal, known at the time as "retail wheeling," an electricity user at point X could contract with a power plant owner at point Y to provide electricity to the plant, and would pay whatever the contract price was to the supplier rather than pay the utility's rates. The utility would be paid only for the service of "wheeling" the power from the supplier rather than the rates it charged normally

Such a concept was promoted not only by the large industrial users of electricity, who would be its direct beneficiaries, but also by free-market enthusiasts and consultants who had developed similar market-based electricity programs in other countries and were eager to perform the same experiment in California. These proponents of restructuring argued that the creation of wholesale and retail markets in electricity, rather than a regulated cost-based system as had previously existed, could deliver better competition and thus lower electricity prices.

The proponents pointed out that thousands and thousands of expert consultant hours had gone into the design of a carefully structured program that could be implemented to provide rate relief for not only large industrial customers, but also for everyone. The marketplace would determine prices—not by the need to service the obligations utilities had incurred for the independent power contracts or for the few (but costly) nuclear power plants that were completed by the late 1970s, and had commenced before energy planning was established on a regular basis in California.

Unfortunately economic theory in this case is quite clear and not subject to the sorts of doubt described previously (as in Chapter 4): a competitive market would require utilities to absorb all of the difference between the costs incurred previously to secure the power (including the high costs of nuclear plants and QF projects) and the presumably lower cost that would be provided in the marketplace. This outcome clearly does not benefit utilities!

In fact, utilities clearly opposed restructuring at first, as did environmentalists and many other stakeholders. Yet notwithstanding those objections, the ideology of economic fundamentalists and the political power of large industrial energy users drove the momentum for electricity restructuring to grow relentlessly.

(An interesting irony is that if environmentalists had not stopped the construction of unnecessary power plants earlier, the political momentum to restructure (caused by high rates) would have been even higher, and the potential damage to the utilities from their inability to pass on fixed costs would have been more dire.)

After the publication of the California Public Utilities' Commission "Blue Book" in 1994, which laid out a detailed proposal for restructuring but did not make any firm decisions, stakeholders began to see restructuring as inevitable. They slowly adjusted their advocacy to the mode of damage control rather than getting what they actually would prefer.

Damage control was necessary because of the way markets work in reality (as opposed to how economic fundamentalists think they work). Markets respond to risk, and risk consists not only of what is expected to happen but also on perceptions of what might happen. Thus the mere fact that the PUC was considering restructuring seriously affected markets. And it affected them all adversely because markets dislike risk.

For efficiency advocates, perceived risk was a problem because, as discussed, some possible outcomes of the "Blue Book" involved regulatory regimes in which spending money on efficiency reduced the competitiveness of the utility that spent it. Utilities responded by slashing their efficiency budgets by half. Because efficiency spending is so cost-effective, this was costing the California economy $1 million a day.

As utilities could not predict their future returns with confidence, their stock prices suffered. And the chosen form of restructuring required utilities to divest themselves of most of their generation assets, clearly something utilities would have preferred to avoid.

So there were manifold reasons why a number of stakeholders had an interest in a prompt even if imperfect resolution of the issue.

What began as a formal proceeding at the PUC quickly turned into a legislative battle, as numerous parties attempted to maximize political advantage. Legislative meetings that involved stakeholders as well as legislators became increasingly intense. Many participants described the process as a "forced march."

In 1997 the California legislature passed restructuring legislation AB 1890, which was based on a grand compromise. Each of the major stakeholder groups had its most important interests protected, while the coalition of large industrial users and economic-theory optimists were able to get their restructuring experiment. Utilities and environmentalists endorsed the bill and consumer organizations refrained from opposition.

Why did the aforementioned bill attract such support? Most stakeholder groups apparently felt that it was a bad idea whose time had come.

Environmentalists felt that restructuring and the creation of open markets would be accomplished either with their support or over their objections. By obtaining a system in which

the utilities could collect funds to run energy-efficiency programs regardless of which electricity supplier would eventually
provide the consumer with electricity, and by restarting state
programs to purchase renewable electricity supplies, the major
source of environmental (and economic) benefits of the status
quo was preserved.

Consumers were able to secure fixed electricity prices for a
certain time into the future and guarantees of rate reductions
to the residential sector.

Utilities were able to assure recovery of stranded costs, at
least up to a fixed point in the future.

So all stakeholders recognized what seemed to be their best
deal, and accepted it. It would be incorrect to say that anyone
except perhaps the original proponents of restructuring was
happy with the result, but almost everyone considered AB 1890
as (at least) the lesser of possible evils.

Clearly the myth that the problem was caused by greedy utilities makes no sense upon close examination. Things obviously
worked out badly for the utilities because they lost significant
amounts of money (and in one case went bankrupt). Yet even in
prospect, the idea of deregulated prices and divestiture of assets
couldn't be beneficial from the utility perspective. Utilities also
would dislike AB 1890 because it left them with the image
that they were responsible for keeping the lights on without the
actual authority to do so as they were unable to purchase new
power plants or make long-term contracts for power.

Proponents of the greedy utilities myth seem to ignore just
how little political power the utilities actually had during the
restructuring experiment. At the time of the crisis itself, when
utilities were buying electricity for $.25 to $.75 per kWh and
more and selling it for $.14 per kWh or less, utilities had so
little political power they could not convince the PUC to raise
rates and keep them out of bankruptcy!

Retrospective reviews of restructuring often blame the state
for its failure to raise rates (discussed below), but despite the
need to protect the financial solvency of the utilities, rates were

increased only after bankruptcy was in progress. This observation thoroughly refutes of the myth of greedy utilities causing
the crisis through their political power.

Myth 4: The Flaws of Restructuring Were Clear to All

The response to this myth about the obviousness of the flaws of
the California restructuring plan shows the seductive power of
economic fundamentalism. Proponents of restructuring, both
in the PUC and in the private sector, spoke at length of the
values of competition and market forces and their ascendance
over monopoly-based decision-making. While consultants who
played the major role in designing the system may have spent
thousands of hours working out all the details, they did so without listening to skeptics. Instead, they were convinced, based on
their work on restructuring in the United Kingdom and other
foreign countries, that they could get it right.

The generic problem the consultants either didn't see or
refused to face is that markets perform differently depending
on the rules. As we have seen in the Chapter 4, there is not
only one choice for the rules that establish an ideal market—
instead there are many remarkably different ways of structuring
a market—and the architects of California's restructuring only
looked at a few.

Many people, including me, pointed out the dangers and
risks of a fully deregulated electricity system that included retail
wheeling. The fundamental physical problem is that electricity delivered through a network of wires is unlike corn or steel
delivered via railroad cars: you can't track the progress of a particular unit of electricity from the supplier to the user. Thus a
power plant at point B that supplies power to a user at point A
cannot be conceptualized as physically shipping the electricity
along the wires from point B to point A.

A better metaphor is a giant bathtub, wherein the user at
point A draws water from the tub and is contracted with the
supplier of a faucet at point B to pour water into the tub. If

everything works as planned, the difference between the two situations isn't terribly critical. However, if supplier B fails to perform as expected, the consequences do not fall merely on user A, but on the system as a whole. One of the prime assumptions needed to make markets work—namely that one consumer's or producer's decision fails to affect anyone else—is wrong. Everyone uses the same wires, so if one consumer or producer creates any electricity shortage, the whole system can black out

A second problem that I warned about, which did actually happen, is that the political system does not allow failure of a critical market, such as electricity. Thus, if industry A contracted with supplier B and suddenly supplier B charged prices ten times as high as had been anticipated, consumer A would simply tell her Congressional representative or state legislator that the plant and its thousand jobs would move out of state if electricity continued to be priced at such an unreasonable rate. Similarly residents of a senior citizens' complex who had contracted with an unreliable supplier of electricity could not be allowed to die of heat exhaustion during the desert summer just because they couldn't afford to pay the electricity rates to run the unexpectedly high-cost air conditioners.

In the case above, market forces would only work up to a point, and then political forces would take over. Because this likely result would be evident to market players, which included both users and generators, consumers would be encouraged to take high risks to save money, knowing that they would be bailed out, and generators would avoid risk (that is, avoid new construction) knowing that they might be unable to recover costs.

So retail wheeling becomes a system to socialize risks and privatize benefits, a system guaranteed to set up the wrong incentives.

I noted that the system of depending on retail wheeling was analogous to the system of regulating savings and loan associations in the 1980s, where a relaxation of controls on how the associations could invest their money coupled with the government's provision of deposit insurance—another system of socialized risks and private rewards—set up misplaced incentives for the associations to take unwarranted risks. This decision to relax controls eventually cost the United States government hundreds of billions of dollars. I also expressed the concern that if we relied solely upon the market to balance supply and demand, it would mean that no one was authorized to spend any money to "keep the lights on." Instead it would be in the interest of all power suppliers to allow a crisis in supply to exist because it would support very high prices, at least for a limited period of time.

The few influential economists who wanted to create "perfect electricity markets" did not sway all consultants. Some consultants noted that a competitive market requires many buyers and many sellers and the transmission of market signals between them. In California's electricity market, even after deregulation, there were and still are a limited number of power plants with well-known capabilities. Each key plant could exert market power during certain peak demand periods, regardless of the plant's ownership. So there truly aren't "many sellers." And given the dominant role of utilities as the buyers of electricity, there are not "many buyers" either.

The fact that not many sellers exist highlights an interesting failure mode: if the number of sellers is limited, it is in the interests of each supplier to create a shortage. If no one invests in new power plants, then the problem will eventually be reflected in high prices, which is to the benefit of the current power plant owners. Owners of existing plants—namely, the companies with the greatest experience and expertise at building and operating plants—could see that they might make more money not building new capacity than building it.

And operating plants would be at no risk of falling behind new competition because building a power plant is a very public process. If an economic incumbent plant owner were to face new competition, the company would know years in advance about the threat.

Critics of restructuring also noted that the market signals

are distorted by market imperfections, such as the inability to meter electricity based on time of day (which is a significant imperfection because the cost of electricity varies radically depending on time of day), and by federal regulations. Therefore the "perfect competitive market" was going to be elusive, and require better market rules and better oversight than was initially implemented in California's rush to embrace so-called competition.

Despite such objections, increasing political momentum was present for restructuring. Different stakeholders in this process had different concerns, all of which would have been compromised by a direct move to so-called free markets.

Consumers were concerned that the best deals would go to the most sophisticated big users, and that residences would be stuck with high and unstable rates. Environmentalists were concerned that the immense progress that the state was making on utility-sponsored efficiency programs would be undercut under restructuring. If utilities' rates were regulated, expenditures on energy efficiency that benefited everyone could be spread over the whole system, which was the ratemaking structure used up until the mid-1990s (and reestablished in the 2000s). But in a competitive environment, a utility that charged extra for efficiency services could be at a competitive disadvantage compared to one that didn't because its rates would be (trivially) higher. (Rates would be higher, but bills would be lower.)

Environmentalists pointed out that the known savings from energy-efficiency programs were much larger than the potential savings that could occur due to competition that forced down the price of energy supply. Environmentalists were also concerned that a state program that encouraged investment in renewable resources would cease, which indeed it did after restructuring became a political likelihood.

Worse yet, many parties recognized that continued uncertainty over restructuring could be worse than an adverse decision. Energy-efficiency advocates saw some $2 million a day in efficiency benefits slipping away until the issue was resolved, and were inclined to accept a flawed solution sooner rather than hoping (probably in vain) to defeat the more seriously flawed proposal later.

Because of the strength of proponents of restructuring, all of these concerned parties were forced to succumb to political pragmatism and spend their political "chips" to take care of their specific concerns while they conceded that restructuring would happen. No one wanted to stand in front of the speeding train of restructuring, even if they thought the train was speeding toward a wreck, which it turned out it was.

In addition, once it became the Legislature's evident goal to restructure the market, in an adversarial manner if consensus could not be found, participants became concerned about where they and their concerns and even their jobs would fit in the new structure. Rather than question the market design and try to identify gaps in its structure, which was being determined by those with political power, technical and regulatory personnel at the time became more focused on meeting the Legislature's 1/1/98 deadline and did not have time to question the regulators' and politicians' rush to deregulate or to correct the flaws in the market design.

Myth 5: Restructuring Did Not Go Far Enough

Perhaps the most pernicious myth of why restructuring didn't work is because it "didn't go far enough." Proponents of this particular version of the myth claim that if high prices had been passed on to retail, everything would have solved itself smoothly.

The facts don't support the assertion that consumers would have reduced their demand enough to prevent high prices. As mentioned, the response of San Diego consumers to a two-fold electricity price increase was only a 2 percent reduction in consumption. Even with the hockey stick price curve shown in Figure 10 in Chapter 5, a 2 percent reduction in demand would have been insufficient to solve the problem. Instead, as illustrated below, the problem would have simply migrated to a different form and a different structure.

In addition, the promoters of this myth seem to have forgotten that passing actual costs on to retail was technically impossible. Electricity use varies from hour to hour. Figure 10 illustrates why electricity price must also vary from hour to hour. This variation can be by as much 5 (or 10) to 1. Yet residential electric meters cannot measure by time of day, nor do most commercial and industrial meters.

While time-of-use meters exist, they are expensive. In the 1990s they were not cost-effective for use in residential or small commercial applications.

When electricity costs are the highest, on a hot summer afternoon, utilities would be unable even technically to pass on costs to retail without years of investment in new meters. The best they could have done was to pass on average costs. This still would not have sent the right price signal to consumers. In the absence of hourly or daily market signals, consumers were unlikely to change consumption patterns, except to scream in agony thirty days later when they received their bills and to look for some entity to blame.

Screams of agony have political as well as economic consequences, and the rate increases would not have been allowed to persist. It is worthwhile to recall that the CPUC and the governor were unwilling to allow rate increases until well into 2001. This is a rational political response to public attitudes: the attitude of "we won't pay these exorbitant prices" took precedence over demand response in San Diego. The attitude "let PG&E and SCE shareholders pay for these outrageous prices" prevailed in the minds of the CPUC and a governor who delayed action, apparently in hopes the crisis would disappear. (Note how this observation refutes Myth 4 that deference to the utilities was the cause of the problem.)

So even if the economists' solution of passing prices on to consumers had been employed, it would not have solved the problem, for several different reasons. First, as noted, it was technically impossible to pass on the real, hourly prices.

Second, the low level of responsiveness to price shows that even if it had been possible to pass on the full real prices, the state still would have had a power shortage and still would have paid excessive prices (although not quite as high as what actually happened).

Third, while the utilities may not have lost all the money, consumers and businesses would have been stuck with much heavier financial burdens than they were. Californians would still have had to absorb most of the $15 billion in extra costs. What effects might have occurred? Because such a scenario didn't occur, it is difficult to predict what would have happened, but the effects could have been even worse. For example, elderly people in hot areas might have been unable to afford air conditioning, and some may have gotten sick or died. Businesses with high electricity bills may have gone bust or moved out of state.

One possibility worth a mention is that the rule of law itself might have broken down with respect to electricity. Economic fundamentalists forget that markets are premised on a number of assumptions. As noted in Chapter 4, these assumptions include the rule of law, or the expectation that deals are real. In this case the "deal" is that consumers will pay their electric bills based on the meter reading and the posted cost of electricity. But if prices go up tenfold this assumption becomes doubtful.

When prices did in fact double in San Diego, consumers were advised simply not to pay. This advice was not given by an anarchist speaking on a street corner; it came from the Chair of the State Senate Energy Committee. How many consumers actually would have followed through and withheld payment for their utility bills if the cost had risen even higher?

Was a doubling of prices—with the risk of higher prices to come—a sustainable situation in our political economy?

Free markets work in political systems only to the extent that the government doesn't step in and overrule them. But governments throughout the world have a pretty consistent track record of overruling free markets when the results become too painful to people. When inflation gets too serious, even Republican and conservative administrations in the United States clamp on price controls. And when utility prices start to go up, governments frequently intervene to shelter consumers from the problem.

For example, when Russia deregulated fuel prices and costs rose to world levels in the 1990s, provincial governments quickly discovered that residents couldn't (or wouldn't) pay their heating bills. In order to avert a total breakdown, the provincial government started to subsidize these bills. Even as late as the year 2000, most consumers' heating bills were subsidized to the tune of 60 to 80 percent by municipal governments, simply to assure that even partial payment was received.

So an attempt to pass price increases on to consumers might not have been practically possible. Either by private action or by political intervention, it is likely that utilities that were still prohibited from entering into long-term contracts would have been unable to collect as much money for power as they had spent.

So the attempt to pass prices on to consumers wouldn't have made much difference in terms of overall cost to the state—it would have failed to lower observed prices significantly. And it also would have failed to address the financial problems that resulted in the bankruptcy or near-bankruptcy of utilities.

Perhaps passing on price increases to consumers would have helped mitigate the crisis. But the mitigation would not have been they way the economic fundamentalists would want us to believe.

First of all, economic fundamentalists would see high prices passed on to the consumer as the normal functioning of competitive forces. While they would perhaps be loath to admit it in public, the theory of ideal free markets would explain that periods of high price were the best possible result: an illustration of how markets balance supply and demand through the price mechanism.

But as a result of all of the failures of the assumptions behind economic fundamentalism, as well as the failures of the market, even this bleak interpretation would be overly optimistic. Power plant owners would see extreme prices as a continuing chance to make windfall profits. The developers of new plants—typically the same companies that were making billions from existing power plants—would not be encouraged to build new capacity in response to high price, they would see such prices as a distraction: an "opportunity" to spend five years constructing a facility that would be finished just too late to cash in on the high prices.

So the market would be slow to address the problem of price. Prices would stay high for years. And when new companies finally added capacity to the system, there would be no reason to believe that they would build the right amount. The result could be a later surplus of power and very low prices—so low that they threatened the financial stability of the suppliers. Unstable suppliers would present yet another set of challenges to California consumers and businesses.

The effect of such continuing high prices would not have been exclusively economic—supply and demand that comes eventually into equilibrium due to price effects. It would have been political: perhaps allowing rate increases to protect the economic integrity of utilities; perhaps regulation of electric prices at wholesale; or perhaps something more difficult to predict. However, the result would not be what is expected by simple economic theory.

Myth 6: The "Genius of the Market" Will Solve Everything

The economic fundamentalists who designed the California system truly believed that the market would solve everything, and stated this belief explicitly. In fact, the very phrase, "the genius of the market" was taken from a statement by the president of the PUC (quoted at length below) as he explained why sole reliance on spot markets was the best policy.

Proponents of California restructuring acted as if the establishment of a system of competition among generators would inevitably bring down costs, and that any more detailed thought was unnecessary and even counter-productive. Market forces, they believed, should handle the financial issues of electricity competition, not special deals between retail utilities and generators. (In this respect, they agreed with the observation I made in Chapter 5 that markets are not truly free when large players can make deals with preferred business partners.)

If customers were interested in long-term stable prices, they could enter into financial contracts for them. Such markets would naturally emerge, according to economic fundamentalism, so those who wanted could purchase exactly the desired amount of stability, and those who wanted to chance it with spot prices could also do so. A spot-price-only system would provide the greatest competition and the greatest choice.

Instead, if we look at the details of real markets rather than ideal markets, we can identify how the system so quickly failed.

How did California's approach to restructuring cause the crisis of 2000? First, during the four-year gap between the first serious consideration of restructuring and the eventual passage of AB 1890, the rules of competition—under which builders of power plants could expect to see a return on investment—were unclear.

Utilities were no longer going to invest in power plants because under any concept of restructuring, they would acquire power from independent producers. Indeed, utilities divested themselves of generation assets.

Independent producers would be unwilling to propose building power plants until they understood the rules for the new market that would be created and could guess how much money they would get for power sales based on supply and demand rather than on cost.

And investors would not finance new power plants until a reliable set of rules for cost recovery was known. So basically the high level of risk that an uncertain system of pricing generated led to a total halt in proposals for building both conventional and renewable power plants.

At the same time, utilities, apprehensive about whether the cost of energy efficiency programs would make them uncompetitive suppliers, and even whether they could continue to collect funds for energy efficiency, reduced the expenditures on their programs by 50 percent after 1994. (Even though AB 1890 provided assurance of continuing funding for energy efficiency, the level that was continued was at the lower level. And this bill failed to pass until 1997.)

So as of 1994 utility investments in energy efficiency were reduced by about 250 to 350 megawatts/yr. Investments in renewable power, which had also ran at almost 200 megawatts/yr, disappeared. And virtually no new conventional power plants were built. Established long-term contracts were allowed to expire, and new ones were not negotiated because wholesale transactions would be accomplished via the Power Exchange. So by 2000 the state was short some 3,000 megawatts from efficiency and renewables alone. Sadly this additional cushion would have prevented most of the problem from occurring.

Interestingly part of the reason for the lack of interest in both renewable and conventional power was the widespread perception at the time when restructuring went through that there was a looming surplus of power and therefore that prices would be too low to recover any money from investments.

Because of widespread perceptions of surplus, and the consequent relatively low prices for electricity during the first couple of years of restructuring (which kept everyone content), insufficient attention was paid to planning for what would happen when the surpluses ended or to identify market conditions that might lead to adverse market outcomes. In fact, the California Energy Commission didn't discover the imminent end of surpluses until 1999, and the problem was then forecast for less than two years in advance.

Not until 1999 did more than a few analysts recognize problems because, with deregulation looming, the utilities had disbanded their long-term planning groups in 1996. If markets rather than regulated utilities would decide how much power is available, why spend money on planning? During the period between 1995 and 1997 when restructuring decisions were being made, neither the CPUC nor the Legislature asked for or required substantive quantitative market forecasts or analyses that used system models to examine how the massive restructuring efforts might work under alternative market conditions and market designs.

Less quantitative modeling of alternative market scenarios and forecasts of market behavior were performed for the multi-billion-dollar restructuring than for almost any single utility rate case. Only the CEC continued to carry out and publish market projections, but these were not extensively debated. The rush to deregulate and the tug-of-war between the CPUC and the legislature and advocacy groups did not rely on careful analysis and prudent planning, so the state's regulators and legislators were surprised when the market malfunctioned.

This was a shame because even a rough cut at cost-benefit analysis would have shown that the whole restructuring enterprise was dubious on the economics. The proponents of restructuring looked at competition as a manifest good—something that is desirable for its own sake. (Actually, even economic fundamentalists would have to admit that competition is beneficial only to the extent that it forces companies to reduce prices and maximize economic efficiency.)

Yet if proponents of restructuring had been required to predict the benefits they thought it would bring, the expected public benefit from increased competition would be limited by the expected economic inefficiency in the existing system of electricity distribution and production. No one tried to quantify how significant these inefficiencies that could be avoided were, but it seemed unreasonable at the time to assume that they could be much more than 10 percent.

In other words, the so-called promise for deregulating electricity could have been no more than about a 10 percent decrease in electricity prices. In contrast, the benefits from utility-sponsored energy-efficiency programs, which the original deregulation proposal put at risk (and which were compromised by 30 to 50 percent in the actual enacted law) were a reduction in electricity costs by more than 10 percent.

So a "deregulatory" solution that increased competition on the supply side and also removed incentives on the demand side would actually cost more than it would save. Yet this analysis was ignored, and the experiment proceeded.

Up to the actual point when the system broke down, the proponents of restructuring were confident that they had gotten it right. Although the system formally depended solely on trades on a short-term spot market, its proponents continued to insist up until the crisis began that "the genius of the marketplace" would create long-term contracts that could intermediate the difference between short-term and long-term markets through national financial markets.

They were told, but ignored, that a competitive bidding process that awarded all successful bidders the market clearing price would enable California's least expensive source of imported power (i.e., Northwest hydro, which was previously priced at split-the-savings, cost-of-service prices) to be priced instead at the marginal bid price, usually from a natural gas plant. Thus, essentially, all Northwest energy took a great leap upward in price, and gave the Northwest and all other out-of-state generators windfall profits.

Many other similar operational and pragmatic details were ignored, as regulators focused instead on economic-theory issues of market structure.

For example, in addressing why it wanted to place total reliance on short-term markets, the PUC stated:

> Many customers may be disinterested in the choice of generation but desire price stability and predictability over a defined period of time. Such customers are free to elect hedging contracts which may be concluded with any individual or entity willing to take the counter-part risk.

In our view parties [who] agree to accept the risk in a hedging contract may have generation facilities or contracted rights to generation but we see no need to restrict their qualification or in any manner make hedging contracts, termed "contracts for differences" in much of the literature, the object of Commission concern. **Both entry into and exit from such a business, as well as the terms of such contracts are left to the genius of the marketplace and the will of market participants.**[4] [Emphasis added]

In sum, both the details and the overall structure of the experiment were considered as based on the highest levels of expert advice and economic wisdom, according to the proponents. Decision-makers ignored the voices of opposition; and warnings were not merely dismissed, but totally ignored.

While many "Monday-morning quarterbacks" have suggested that the restructuring proposal was poorly thought out and incomplete, its proponents, including mainly the CPUC, believed up until the crisis hit that they had gotten it right—that they had unleashed market forces that would lead to greater competition and lower costs. And if one believes the economic fundamentalist view of the world, they did pretty much get it right.

The problems that occurred illustrate the failures of economic fundamentalism. Let's begin with the PUC statement above. The PUC believed that financial markets would provide the price security that was desired in the marketplace, so the reliance on spot markets was best. The problem was that while financial markets could have provided this stabilizing influence, in fact they didn't. This is the problem referred to in Chapter 4 of "the lack of a well-defined equilibrium finding mechanism"—even if a set of contracts for differences is the Pareto-optimum answer, no assurance exists that markets will get us there.

Notes

Introduction

1. T. Kubo, H. Sachs, S. Nadel. "Opportunities for New Appliance and Equipment Efficiency Standards: Energy and Economic Savings Beyond Current Standards Programs. American Council for an Energy Efficient Economy," ACEEE Report #A016, 2001.

2. "The Benefits and Costs of the Clean Air Act 1990–2010." U.S. Environmental Protection Agency. November 1999. (See Tables 8-2, 8-3, and 8-4 for summary results.)

Chapter One

1. Data on emissions are available from the U.S. Environmental Protection Agency's "National Emissions Inventory" and are summarized in "National Air Pollutant Emissions Trends 1970–98."

2. Thomas L. Friedman. "Too Much Pork and Too Little Sugar." Op-ed in the *New York Times*, August 5, 2005.

3. We can argue that higher prices for energy will help induce these behaviors, yet higher prices are a politically infeasible solution in the United States. And even if energy prices could be changed, the amount of energy-conserving behavior that would be induced is not terribly large. The potential for price-driven behavior change is less than most economists predict because conventional models assume that price is the only force causing reductions in energy use and therefore much of the savings that are actually caused by standards are incorrectly attributed to price.

4. The Snowe (R-ME)/Feinstein (D-CA) EFFECTER Act, S. 680

5. "Informing Regulatory Decisions: 2003 Report to Congress on the Costs and Benefits of Federal Regulations and Unfunded Mandates on State, Local, and Tribal Entities." U.S. Office of Management and Budget, Office of Information and Regulatory Affairs, 2003.

6. Carrier Corporation press release, 27 August 2005.

7. Dan Sperling et al., "The Price of Regulation," *Access,* Number 25, Fall 2004.

8. One of the more recent studies evaluated the results of $1.4 billion dollars of utility investment in efficiency during 2000–04, a very substantial sum. The study indicated that the cost of saved electricity was 2.9 cents per kWh. This is much lower than the cost of supplying that kWh would have been, namely, 5.8 cents for off-peak energy (winter, summer nights, and weekends), 16.7 cents for on-peak energy (summer weekday noon to 7 p.m.), and 11.7 cents for shoulder hours (all other hours). C. Rogers, M. Messenger, S. Bender. Funding and Savings for Energy Efficiency Programs for Program Years 2000–2004. California Energy Commission. CEC-400-2005-042, 2005.

Chapter Two

1. Vice President Cheney claims in his report, "National Energy Policy," that one-half to two-thirds of the 42-percent improvement is a result of energy efficiency—the rest was due to higher prices or changing industrial mix.

2. M. Ross, R. Socolow et al. *Efficient Use of Energy: A Physics Perspective.* American Physical Society, January 1975.

3. Amory B. Lovins. "Energy Strategy: The Road Not Taken." *Foreign Affairs,* in October 1976.

4. Note that some of these data are reconstructed and therefore approximate. The numbers for 1972 are relatively accurate because they are based on actual test results for each model produced, performed by manufacturers, weighted by the actual number of sales of each individual model. Thus the only source of error is the difference between the test procedure results and the results in the field, which a number of studies over the years have shown as relatively small. The results for earlier years are inferred based on measurements of refrigerator energy consumption in the 1947–1950 period and mid-1970s measurements of the energy consumption of a few such products that still were operational by that time, establishing the original number at about 350 kWh/yr.

5. Most directly, an electric heater accomplishes the defrosting, which uses energy. Second, with manual defrosting the cooling coils are located in the refrigerated space, and can cool the space by radiation and natural convection without the need for additional energy use. But to defrost, the cold coils must be outside the refrigerated space (otherwise the defrosting process would overheat the stored food). Thus the refrigerator requires a fan to transfer cooled air to the space. Also, because of losses in transferring cooled air, the temperature of the coils must be a little colder,

which makes the refrigeration system less efficient. Then all the energy used by the defroster and the fan turns into heat inside the box, which must be removed by running the refrigerator strenuously. In addition, the air movement the fan generates can cause air or heat leakage through the gaskets. As suspected, all of these problems compound each other; every time the system adds heat or loses cool air, it must run harder.

6. Side-door refrigerators use more energy because the length of the door gaskets is greater, and door gaskets are a disproportionate source of heat gain (both through air leakage and because they are more difficult to insulate). Through-the-door service increases energy because the box that houses the water and ice dispenser cannot be insulated to the same thickness as the surrounding walls, and because the water pipes conduct heat. However, these losses more than compensate for consumers' reduction in door openings.

7. RAND Corporation performed a study for the California Legislature in the early 1970s that projected the need within the foreseeable future for a large, new power plant sited every few miles along the whole coastline from Mexico to Oregon. Because most of the coastline is public beaches, oceanfront luxury properties, or relatively undisturbed wild areas, most voters found this prospect unattractive.

8. The formal participants in SERP were the utilities that funded it: Pacific Gas and Electric, Southern California Edison, the Sacramento Municipal Utility District, San Diego Gas and Electric, the Los Angeles Department of Water and Power, the Northern California Power Agency, the Bonneville Power Administration, PacifiCorp, Portland General Electric, Arizona Public Service, Long Island Lighting Company, New England Electric, Public Service Gas and Electric, Commonwealth Electric, Western Massachusetts Electric, Central Maine Power, Jersey Central Power, Atlantic City, Baltimore Gas, Madison Gas, Northern States Power (both Minnesota and Wisconsin), Superior Water Power and Light, Wisconsin Electric, Wisconsin Power and Light, and Wisconsin Public Service.

In addition, the program was developed with the extensive participation of the U.S. Environmental Protection Agency, the Washington State Energy Office, the Natural Resources Defense Council, and the American Council for an Energy Efficient Economy.

9. Whirlpool's winning bid relied entirely on energy savings and cost for its high ranking—it did not offer additional ozone-depletion reductions.

10. The average energy consumption of an automatic-defrost refrigerator in 1972 was 2,127 kWh/yr.

11. D. Goldstein. Preventing Wasted Light. *The Construction Specifier.*

October 1984; also published in Energy Technology IX. Rockville, MD: Government Institutes, Inc., 1984.

12. Apparently it's also more fun to drive. All of the author's friends who own the Toyota Prius insist that they enjoy it for its responsiveness even more than for its fuel economy.

13. If solar heat gains are large, increasing insulation will actually worsen the amount by which interior temperatures in the car exceed ambient temperatures when the car is parked in a sunny parking lot.

14. J. Holtzclaw et al. "Location Efficiency: Neighborhood and Socio-Economic Characteristics Determine Auto Ownership and Use—Studies in Chicago, Los Angeles, and San Francisco." *Transportation Planning and Technology Journal.* Volume 25, Number 1 (March 2002).

15. M. J. Burer, D. B. Goldstein, J. Holtzclaw. "Location Efficiency as the Missing Piece of the Energy Puzzle: How Smart Growth Can Unlock Trillion Dollar Consumer Cost Savings." *Proceedings of the 2004 Summer Study on Energy Efficiency in Buildings.* American Council for an Energy Efficient Economy, Washington, D.C., August 2004.

16. A family that spends $8,000 a year for the life of a thirty-year mortgage will pay $240,000. (This number includes the assumption that future savings must be discounted, but also that future driving costs will be higher even after adjusting for inflation. It is also a little low because the driving costs referred to the year 1998 or so, thus the real costs are a little higher now.) The average cost of a new home was $204,000 in July 2005, according to the U.S. Department of Commerce, as reported on CNN.com on August 24, 2005.

17. M. Porter and C. van der Linde. "Toward a New Conception of the Environment-Competitiveness Relationship." *Journal of Economic Perspectives.* Volume 9, Number 4, 1995.

18. Joseph J. Romm. *Cool Companies: How the Best Businesses Boost Profits and Productivity by Cutting Greenhouse Gas Emissions.* Washington, D.C., Island Press, 1999.

19. For the most recent example, see the Northwest Power Planning Council's 2005 document, "The Fifth Northwest Electric Power and Conservation Plan," which finds that in most scenarios efficiency is virtually the only new resource needed for the rest of the decade.

20. California Energy Commission, "Implementing California's Loading Order for Electricity Resources," Staff Report, Publication CEC-400-2005-043, July 2005, Figure E-1, p. E-5.

21. The fact that the national government was not truly trying is illustrated by the fact that the Department of Energy has missed over twenty-five of its statutory deadlines to set appliance and equipment efficiency standards, and has yet to set a single standard that is discretionary, despite

broad recognition that the appliance and efficiency standards program is the nation's most effective energy policy. The Department of Energy's Office of Energy Efficiency and Renewable Energy reveal similar results. These budgets, which are mostly for technology research and development, were expanded in the wake of the energy crises of the 1970s, but cut back drastically in the 1980s. In real dollars, they are still substantially below the level of the 1970s. The Department spends about $1 on efficiency research and development for every $1000 spent on energy; in comparison, most companies spend about 1 percent of gross revenues on research and development, and fast-growing high-tech companies spend over 5 percent.

22.Typical space-heating energy use for a city with Hood River's climate was 13,000 kWh per year. After the project, measured usage was less than 5,000 kWh per year. This change was due to a number of economic and energy factors in addition to the efficiency measures, including heavy use of wood heating and an economic downturn that caused consumers to heat their homes less. Eric Hirst. "Cooperation and Community Conservation: Final Report, Hood River Conservation Project." Oak Ridge National Laboratory, DOE/BP-11287-18.

Chapter Three

1. "Low-e" is an abbreviation for low emissivity; a low-emissivity window reflects infrared radiation. (Low emissivity and low reflectivity always go together.) When the emissivity is low for thermal radiation—the part of the infrared spectrum that radiates room temperature heat—the coating makes the window a better insulator. The low emissivity means that one pane of the window can't easily radiate heat to the other. When the emissivity is low for the higher-frequency infrared radiation that is found in sunlight, the window also reflects this infrared solar energy.

2. "Smart Growth: More Choices for our Families." Surface Transportation Policy Project, Washington, D.C., 2002.

3 .Linda Greer, "Anatomy of a Successful Pollution Reduction Project." *Environmental Science & Technology* (June 1, 2000) [pp. 66–73].

4. Birgitta Forsberg. "Component Compliance," quoting Francois Gauthier, vice president of Coherent. *San Francisco Chronicle*, pages E1 and E7, 27 February 2005.

5. Richard L. Meier. *Planning for an Urban World: The Design of Resource-Conserving Cities.* Cambridge, MA: MIT Press, 1975.

6. Curtis Moore and Alan Miller, *Green Gold: Japan, Germany, the United States, and the Race for Environmental Technology.* Boston, MA:

Beacon Press, 1994.

7. Randall Dodd. "The Wealth Curse from Natural Resources." Washington, DC. Financial Policy Forum, Derivatives Study Center, Special Policy Brief #17, 2004.

8. Richard Florida. *The Rise of the Creative Class*. New York, Basic Books, 2002.

9. ———. *Cities and the Creative Class*. New York: Rutledge, 2005, pp. 5667.

10. Ibid.

Chapter Four

1. For example, two widely used textbooks on economics, one of them authored by two of the most esteemed economists in America, spend hundreds of pages to describe what economics analyzes and what sort of conclusions economics reaches with scarcely even a mention of what assumptions underlie the conclusions! Both textbooks—Paul A. Samuelson and William D. Nordhaus. *Economics (12th Edition)*. New York: McGraw-Hill, 1985, and Campbell McConnell. *Economics*. New York: McGraw-Hill, 1975—explicitly list as assumptions only the third and fourth ones listed here in the text, although they discuss the first two as beliefs that are taken for granted.

2. Stephen J. DeCanio. *Economic Models of Climate Change: A Critique*. New York: Palgrave Macmillan, 2003, p. 9, citation of definition stated by noted economist Andreu Mas-Colell.

3. Many of my neighbors buy compact fluorescent lamps in part due to programs that my utility has offered over the last decade to promote its sale and assure their quality.

4. Paul Ormerod, "Butterfly Economics." New York: Pantheon Books, 1998.

5. John Kay. *Culture and Prosperity*. New York: HarperCollins. 2004.. p. 319; emphasis added.

6. DeCanio, p. 47.

7. ———.

8. See, for example, Leviticus 19:35-36; Deuteronomy 20:14 and 26:13-15; Proverbs 11:1 and 20:10; Amos 8:4-6, Micah 6:4-6; Ezekiel 45:10-12. The Qur'an also provides guidance on the importance of honest weights and measures, for example, in Surah 17: 35.

9. Ross Cheit. *Setting Safety Standards: Regulation in the Public and Private Sectors*. (University of California Press, 1990).

10. Personal communication, Maureen Breitenberg, conveyed via John Talbott, October 2004.

11. As Joseph Stiglitz explains in a book review of *The Moral Consequence of Economic Growth*, by Benjamin Friedman in *Foreign Affairs*

("The Ethical Economist," *Foreign Affairs*, November/December 2005, pp. 128–134, Vol. 84, No. 6 (Nov./Dec. 2005), p. 133), "American economists tend to have a strong aversion to advocating government intervention. Their basic presumption is often that markets generally work by themselves…government economic policy, the thinking goes, should include only minimal intervention to ensure economic efficiency. The intellectual foundations for this presumption are weak…. Economies are not efficient on their own. This recognition inevitably leads to the conclusion that there is a potentially significant role for government."

Chapter Five

1. In economic terms this is referred to as an imbalance between supply and demand. In this situation the demand for electricity in California exceeded the supply available at reasonable prices within the western United States, which caused hourly prices in California to go through the roof on a number of occasions and led to curtailments when not enough supply was available to meet demand. Shortfalls in hydro-generation, the actions of some market participants to manipulate the market, and other causes exacerbated the situation.

2. The "California energy crisis" is generally considered to have begun in earnest in May 2000. The CPUC responded and raised rates somewhat in January 2001 and gave authority to the Department of Water Resources to purchase power in lieu of the financially strapped utilities. Due to its inability to raise rates and its necessity to continue to purchase power at market prices, PG&E declared bankruptcy on April 6, 2001. Californians will pay off the costs incurred throughout the crisis over the next decade.

3. Energy conservation refers to changes in behavior that may slightly cut back on energy services while saving energy and are, at any rate, short-term actions; energy efficiency refers to long-term activities that maintain or enhance energy services while saving energy.

4. While economic fundamentalists would argue that deregulation did not go far enough and that as a result prices were higher than they should have been in 2000, the remarkably low responsiveness to price in San Diego shows that even if prices were passed on to retail to the extent technologically possible, most of the damage would still have resulted.

5. These reductions compare the peak load of June 2001 with June 2000, July 2001 with July 2000, et cetera, without any adjustments.

Chapter Six

1. A clear explanation of these factors and how they account for market failures is presented in Lorraine Lundquist's paper, "Risk and Loss Aversion: Implications for Environmental Policy." Lundquist is a gradu-

ate of the Energy and Resources Group at the University of California at Berkeley.

2. R.H. Thaler. "Toward a Positive Theory of Consumer Choice." *Journal of Economic Behavior and Organization.* 1: 39-60, 1980.

3. D. Kahneman and A. Tversky. "Prospect Theory: An Analysis of Decision under Risk." *Econometrica* 47: 262-93, 1979.

Chapter Seven

1. *Washington Post*, 20 February 2003.

2. Antonio Regalado. "Global Warming: In Climate Debate, The 'Hockey Stick' Leads to a Face-Off." *Wall Street Journal.* 14 Feb 2005, p. 1 and A13.

3. J. Kay, p. 346.

4. See Ormerud.

5. "The Mercury Scare." *The Wall Street Journal.* April 8, 2004. p. A16.

6. Indeed, the *Wall Street Journal* published a news article on February 20, 2003, "Mercury Threat to Kids Rising, Unreleased EPA Report Warns" by John J. Fialka, that demonstrates that their editorial writers knew, or easily could have known, the facts on mercury exposure from fish consumption.

7. Some individual companies do not agree with these positions and present a more cautionary approach to the problems of climate change in their public statements.

8. Some prominent Republican senators and Congressmembers, as well as governors, have taken leadership positions in support to solve the problem of climate change.

Chapter Ten

1. A large number of books and articles documenting the large efficiency potential that can be realized at no cost has been published over the past thirty years. The following list, in chronological order, is a selection of some of the most convincing:

S. D. Freeman et al. *A Time To Choose*. Cambridge, MA: Ballinger Publishing, 1974.

Ahern, Doctor et al. "Energy Alternatives for California: Paths to the Future," RAND Corporation, R-1793-CSA/RF, 1975.

Lovins, A. and H. *Soft Energy Paths.*

P. Craig, D. Goldstein, R. Kukulka, A. Rosenfeld. "Energy Extension for California: Context and Potential." Proceedings of the 1976 Summer Workshop on an Energy Extension Service. Lawrence Berkeley Laboratory, LBL-5236, 1977.

R.Cavanagh et al. "Choosing an Electrical Energy Future for the

Pacific Northwest: An Alternative Scenario." U.S. Department of Energy. DOE/CS/10045-T1, 1980

L. King, D. B. Goldstein, et al., "Moving California Toward a Renewable Energy Future," Natural Resources Defense Council, San Francisco, 1980.

Solar Energy Research Institute: "A New Prosperity: Building a Sustainable Energy Future—The SERI Solar/Conservation Study." Handover, MA: Brickhouse Publishing 1981.

D. B. Goldstein, M. Gardner, et al. "A Model Electric Power and Conservation Plan for the Pacific Northwest," Northwest Conservation Act Coalition, 1982.

Northwest Conservation and Electric Power Plan Northwest Power Planning Council, Portland, OR, 1986.

J. Goldemberg et al. *Energy for a Sustainable World* World Resources Institute, Washington, D.C., 1987.

"California's Energy Outlook, 1987 Biennial Report" and "1987 Conservation Report." California Energy Commission.

A. Meyer, H. Geller, D. Lashof, M. B. Zimmerman, P.M. Miller, D. B. Goldstein et al. *America's Energy Choices*, Union of Concerned Scientists, Cambridge, MA (1991).

Energy Efficiency Report California Energy Commission, 1993.

S. Bernow et al. "Energy Innovations: A Prosperous Path to a Clean Environment" ASE, ACEEE, NRDC, Tellus Institute, UCS, 1997.

Inter-Laboratory Working Group on Energy Efficient and Low-Carbon Technologies. "Potential Impacts of Energy Efficient and Low-Carbon Technologies by 2010 and Beyond." U.S. Department of Energy, Sept. 1997.

A. H. Rosenfeld and D. Hafemeister. "Energy Efficient Buildings." *Scientific American*, April 1998.

R. Watson. *Oil and Conservation Resources Fact Sheet: A Least-Cost Planning Perspective*. NRDC, San Francisco, 1998.

Douglas H. Ogden. "Boosting Prosperity: Reducing the Threat of Global Climate Change Through Sustainable Energy Investment." American Council for an Energy Efficient Economy, ACEEE Report Number E963, 1995.

Howard Geller, Stephen Bernow, and William Dougherty. "Meeting America's Kyoto Protocol Target: Policies and Impacts. American Council for an Energy Efficient Economy." ACEEE Report # E993, 1999.

Howard Geller, Steven Nadel, R. Neal Elliott, Martin Thomas, and John DeCicco. "Approaching the Kyoto Targets: Five Key Strategies for the United States. American Council for an Energy Efficient Economy." ACEEE Report # E981, 1998.

Inter-Laboratory Working Group. *Scenarios for a Clean Energy Future.* Oak Ridge National Laboratory and Lawrence Berkeley National Laboratory, 2000.

"Cutting Carbon Emissions at a Profit." F. Krause, International Project for Sustainable Energy Paths, IPSEP, 2001.

H. Geller. *Energy Revolution: Policies for a Sustainable Future.* Island Press, 2003.

2. Perhaps the most obvious example of the studies' excessive conservatism was that it projected a 20-percent improvement in air-conditioner efficiency might be obtained in the "high technology," more aggressive scenario. In fact, a 30-percent improvement in air-conditioner efficiency will be required as a minimum standard effective in 2006, and in August 2005 the U.S. Congress established tax incentives for efficiency levels 15 percent higher than that. Many utilities currently pay incentives for these levels, which are 50 percent higher efficiency than the base case. So one critical example, *we already see the real world doing better than the most aggressive scenario modeled by the DOE study.*

3. It is actually accepted in many states, including two large states with Republican administrations.

4. Note that this argument only applies to developing countries because the nation's main economic competitors in Europe and Japan will also be subject to climate restrictions.

5. F. Krause. "Cutting Carbon Emissions at a Profit." International Project on Sustainable Energy Paths, http://IPSEP.org, 2001.

6. Luigi Zingales and Raghuram Ragan. "Saving Capitalism from the Capitalist: Unleashing the Power of Financial Markets to Create Wealth and Spread Opportunity." New York: Crown Business, 2003.

Chapter Eleven

1. D. Driesen. "Is Emissions Trading an Economic Incentive Program?: Replacing the Command and Control/Economic Incentive Dichotomy." 55, Wash. & Lee L Rev. 289, 296–304 (1998).

2. See Driesen and also M. Taylor, E. Rubin, and D. Hounshell, "Regulation as the Mother of Invention: The Case of SO2 Control." 27 Law and Policy 348, 370, 2005.

3. Richard Duke and Daniel Kammen. The Economics of Energy Market Transformation Programs. *The Energy Journal,* 20(4):15–64.

Chapter Twelve

1. Robert F. Kennedy Jr. argues "free-market capitalism is the best thing that could happen to our economy, our country" in *Crimes Against Nature.* New York: HarperCollins, 2004, p. 190.

Appendix A

1. Consumption data are from the California Energy Commission website: http://energy.ca.gov.

2. Christopher Wilkins and M. H. Hosni. Heat Gain from Office Equipment. *ASHRAE Journal*, June 2000, pp. 33–43. For general office equipment actual average power was most accurately estimated as 25 percent of nameplate power. For computers, average power was 15 percent of rated power, without even considering the sleep mode.

3. The FERC decision expressed "grave concern about the need for this capacity" and concluded that California utilities could not lawfully be required to execute the purchase contracts. Federal Energy Regulatory Commission, Order on Petitions for Enforcement Action Pursuant to Section 210(h) of PURPA, Docket No. EL95-16-00 (February 23, 1995), pp. 26–27.

4. Decision 95-12-063 (December 20, 1995) as modified by Decision 96-01-009 (January 10, 1996), at p. 8.

Bibliography

A Retrospective Examination of Long-Term Energy Forecasts for the United States.

Ackerman, Bruce A. and Richard B. Stewart, "Reforming Environmental Law: The Democratic Case for Market Incentives." 13 *Col. J. Envtl. L.* 171 (1988).

Anderson, Steve. *Industry Genius.*

Annual Review of Energy and the Environment November 2002, Vol. 27: 83–118. Download at: http://arjournals.annualreviews. org/toc/energy/27/1;jsessionid=jp828tLivKzd

Bachrach, D. *Energy Efficiency Leadership in California: Preventing the Next Crisis.* 2003.

Bauman, Yoram. "The Effects of Environmental Policy on Technology Change in Pollution Control." Ph.D. thesis submitted to the University of Washington, Department of Economics, 2003.

————. Paradigms and the Porter Hypothesis. Download at: http://www.catobooks.org/pubs/regulation/regv23n1/shaw.pdf

"The Benefits and Costs of the Clean Air Act 1990–2010." U.S. Environmental Protection Agency. November 1999.

Benfield, Kaid, F. Matthew, D. Raimi, and Donald D. T. Chen. *Once There Were Greenfields.* New York. Natural Resources Defense Council, 1999.

————, Jutka Terris, and Nancy Vorsanger. *Solving Sprawl: Models of Smart Growth Across America.* New York. Natural Resources Defense Council, 2001.

Berger, John J. *Charging Ahead.* New York: Henry Holt and Co., 1997.

Berman, S.M., R. Clear, D.B. Goldstein et al. "Electrical Energy Consumption in California: Data Collection and Analysis," Lawrence Berkeley Laboratory, UCID 3847, 1976.

California Energy Commission. 1992–1993 California Energy Plan

Canan, Penelope and Nancy Richman. "Ozone Connection: Expert .Networks in Global Environmental Governance." University of Denver. Greenleaf, 2002.

Cannon, Fred. Bank of America economist, cited in the *San Francisco Chronicle,* 24 June 1993, "Why going Green Pays Off," by Kenneth Howe.

Cavanagh, R. C., D. B. Goldstein, and R. K. Watson. "One Last Chance for a National Energy Policy," in D. E. Abrahamson, ed., *The Challenge of Global Warming,* Washington, DC: Island Press, 1989.

Clark, Woodrow W. II and Ted K. Bradshaw: *Agile Energy Systems: Global Lessons from the California Energy Crisis.* Oxford, UK: Elsevier Ltd., 2004.

"Clean Energy Futures," Department of Energy.

Cleveland, Cutlar J., "Markets Failures in Energy Markets," *Encyclopedia of Energy,* Elsevier Science.

Cook, Elizabeth, ed. "Ozone Protection in the United States: Elements of Success." World Resources Institute.

Craig, Paul P., Ashok Gadgil, and Jonathan G. Koomey. *What Can History Teach Us?*

Cutting Carbon Emissions at a Profit F. Krause, IPSEP, 2001. Download at: http://ipsep.org/

Dittmar, Hank and Gloria Ohland, ed. *The New Transit Town: Best Practices in Transit-Oriented Development.* Washington, DC: Island Press, 2004.

DeCanio, Stephen J. *Economic Models of Climate Change: A Critique.* New York: Palgrave Macmillan, 2003.

Dodd, Randall. "The Wealth Curse from Natural Resources." Washington, DC. Financial Policy Forum, Derivatives Study Center, Special Policy Brief #17, 2004.

DOE Authorization Act.

Driesen, David M. "Markets are not Magic." *The Environmental Journal.* November/December 2003.

"Energy Efficiency Report, California Energy Commission, October 1990." *Energy Innovations* (Tellus Institute: Boston, MA, 1997). Download at: http://www.tellus.org/ei/

Energy Policy and Conservation Act of 1975 (EPCA), 114 Stat. 129, 49 US Code 32901.

"Environmental Law: The Democratic Case for Market Incentives," 13 Col. F.

Fernstrom, G., D. Goldstein, M. L'Ecuyer, S. Nadel, and H. M. Sachs. "Super Appliance Rebates (Golden Carrots) for More Efficient

Appliances: New Incentives for Technological Advances" in Proceedings of the 42nd Annual International Appliance Technical Conference, University of Wisconsin, Madison, WI, May 21–22, 1991, pp. 519–531.

Fialka, John J. "Mercury Threat to Kids Rising, Unreleased EPA Report Warns." *The Wall Street Journal*, February 20, 2003.

Florida, Richard. *Cities and the Creative Class.* New York: Rutledge, 2005.

———. *The Rise of the Creative Class.* New York: Basic Books, 2002.

Geller, Howard. "Energy Revolution: Policies for a Sustainable Future."

———, Stephen Bernow, and William Dougherty. "Meeting America's Kyoto Protocol Target: Policies and Impacts." American Council for an Energy Efficient Economy, ACEEE Report # E993, 1999.

———, Steven Nadel, R. Neal Elliott, Martin Thomas, and John DeCicco. "Approaching the Kyoto Targets: Five Key Strategies for the United States." American Council for an Energy Efficient Economy, ACEEE Report # E981, 1998.

Goldstein, D. B. and H. S. Geller. "Equipment Efficiency Standards: Mitigating Global Climate Change at a Profit." *Physics & Society*, Vol. 28, No. 2, April 1999. (Presented at the 1998 conference of the American Physical Society, Columbus, OH, 14 April 1998.)

——— and M. G. Hoffman. "Forecasting an Increasing Role for Energy Efficiency in Meeting Global Climate Goals." Proceedings of the 2004 Summer Study on Energy Efficiency in Buildings, American Council for an Energy Efficient Economy, Washington, DC, August 2004.

——— and S. Nadel. "Appliance and Equipment Efficiency Standards: History, Impacts, Current Status, and Future Directions." Proceedings of the 1996 ACEEE Summer Study on Energy Efficiency in Buildings, American Council for an Energy Efficient Economy, Washington, DC, 1996.

Goldstein, D. B. "Efficient Refrigerators in Japan: A Comparative Survey of American & Japanese Trends Toward Energy Conserving Refrigerators." *Doing Better: Setting an Agenda for the Second Decade,* American Council for an Energy Efficient Economy, Washington, DC, 1984.

———. "Market Transformations to Super Efficient Products: The Emergence of Partnership Approaches." Proceedings of the 1994 ACEEE Summer Study on Energy Efficiency in Buildings, Vol. 6, American Council for an Energy Efficient Economy, Washington, DC, 1994.

———. "Promoting Energy Efficiency in the Utility Sector through

Coordinated Regulations and Incentives." Physics and Society, Vol. 23 #2, April 1994, presented at the Joint Meeting of the American Physical Society and the American Association of Physics Teachers, Washington, DC, 12–15 April 1993.

———. "Preventing Wasted Light." The Construction Specifier. October 1984; also published in Energy Technology IX. Rockville, MD: Government Institutes Inc., 1984.

———. "Energy Efficiency Opportunities: Why Aren't They Realized in the Marketplace?" NAS presentation October 10, 2001.

Greer, Linda. "Anatomy of a Successful Pollution Reduction Project," *Environmental Science & Technology* June 1, 2000. [pp. 66–73]

Harrington, Winston et al. "On the Accuracy of Regulatory Cost Estimates" Washington, DC: Resources for the Future, January 1999.

Heschong, L. "Daylight and Retail Sales." California Energy Commission P500-03-082-A-5, October 2003.

———. "Daylighting in Schools: Reanalysis Report." California Energy Commission P500-03-082-A-3, October 2003.

———. "Windows and Classrooms: A Study of Student Performance and the Indoor Environment." California Energy Commission P500-03-082-A-7, October 2003.

———. "Windows and Offices: A Study of Office Worker Performance and the Indoor Environment." California Energy Commission P500-03-082-A-9, October 2003.

Holtzclaw, John et al. "Location Efficiency: Neighborhood and Socio-Economic Characteristics Determine Auto Ownership and Use—Studies in Chicago, Los Angeles and San Francisco." *Transportation Planning and Technology Journal*, Volume 25, Number 1, March 2002.

"Integrated Energy Policy Report." California Energy Commission, 2003, and 2004 update. Download at: http://www.energy.ca.gov/reports/100-03-019F.pdf

Kammen, D. M., Shlyakhter, A. I., Broido, C., and Wilson, R. (1994). "Quantifying credibility of energy projections from trends in past data: the U. S. energy sector." *Energy Policy*, 22, 119–131.

Kay, John. *Culture and Prosperity*. New York: HarperCollins, 2004.

Kennedy, Robert F. Jr. *Crimes Against Nature*. New York: HarperCollins, 2004.

Krause, Florentin. "Cutting Carbon Emissions at a Profit, Impacts on U.S. Competitiveness and Jobs," *Contemporary Economic Policy*, Volume 21, No. 21. January 2003.

———. "Cutting Carbon Emissions at a Profit, Opportunities for the U.S." *Contemporary Economic Policy*, Volume 20, No. 4, October 2002.

Kubo, T., H. Sachs, and S. Nadel. "Opportunities for New Appliance and Equipment Efficiency Standards: Energy And Economic Savings Beyond Current Standards Programs. American Council for an Energy Efficient Economy." ACEEE Report #A016, 2001.

Lanoie, Paul, Michel Patry, and Richard LaJeunesse. "Environmental Regulation and Productivity: New Findings on the Porter Hypothesis." Ecole des Hautes Etudes, Montreal, 2001.

Lovins, Amory B. "Energy Strategy: The Road Not Taken?" *Foreign Affairs,* 55:1, October 1976.

Matrosov, Yu. Current status of legal base for buildings energy efficiency in Russia. (75 Kb) Bulletin CENEf: Energy Efficiency, #23, M., 2001.

———. "Energy Efficiency Improvement in Russian Dwelling Houses over the Past Decade." (377 Kb) Czestochowa Technical University, International Conference of Science and Technics, "Problems of contemporary energy saving housing structures focused on optimized use of energetic potential," *Czestochowa* 2003, Poland, pp. 179–192.

———, M. Chao, and D. Goldstein. "Development, Review, and Implementation of Building Energy Codes in Russia: History, Process, and Stakeholder Roles." (51 Kb) 2000 Summer Study Proceedings ACEEE, USA, pp. 9.275–9.286.

———, I. Butovsky, and D. Goldstein. "The Experience of developing the Building Energy Requirements for the Countries with Different Climatic Conditions." (36 Kb) Proceedings of the International Building Physics Conference, Eindhoven, Netherlands, 2000, pp. 613–620.

McConnell, Campbell. *Economics.* New York: McGraw-Hill, 1975.

McDonough, William and Michael Braungart. "The Next Industrial Revolution," in *The Atlantic Monthly,* October 1998. [pp. 28–37]

"The Mercury Scare." *The Wall Street Journal.* April 8, 2004, p. A16.

Meier, Richard L. "Planning for an Urban World: The Design of Resource-Conserving Cities."

Meier, Steven M., "Environmentalism and Economic Prosperity: Testing the Environmental Impact Hypothesis." Massachusetts Institute of Technology.

Meyer, A., H. Geller, D. Lashof, M. B. Zimmerman, P. M. Miller, D. B. Goldstein et al. "America's Energy Choices." Union of Concerned Scientists, Cambridge, MA, 1991.

Moore, Curtis and Alan Miller. *Green Gold: Japan, Germany, the United States, and the Race for Environmental Technology.* Boston: Beacon Press, 1994.

Murray, I. "This Christmas a Red-Green Split?" See: http://www.cei.

org/dyn/view_expert.cfm?expert=227

Nadel, S. and D. Goldstein. "Appliance and equipment efficiency standards." Proceedings of 1998, a CEEE Summer Study of Energy Efficiency in Buildings.

Nadel, Steven, Jennifer Thorne, Harvey Sachs, Bill Prindle, and R. Neal Elliot. "Market Transformation: Substantial Progress from a Decade of Work." American Council for an Energy-Efficient Economy April 2003, at http://aceee.org

National Energy Policy Development Group Report (2001) (Bush/Cheney). Download at: http://www.whitehouse.gov/energy/

Northwest Conservation and Electric Power Plan. Northwest Power Planning Council, 1986.

NRDC: "Slower, Costlier and Dirtier: A Critique of the Bush Energy Plan." 2001.

Ogden, Douglas H. "Boosting Prosperity: Reducing the Threat of Global Climate Change Through Sustainable Energy Investment." American Council for an Energy Efficient Economy, a CEEE Report Number E963, 1995.

Ormerod, Paul. *Butterfly Economics*. New York: Pantheon Books, 1998.

Palmer, Oates, and Portney. "Tightening Environmental Standards, the Benefits/Cost or the No Cost Paradigm." *Journal of Economic Perspectives,* Volume 9, No. 4, Fall 1995.

Porter, Michael E. and Claas van der Linde. "Toward a New Conception of the Environment-Competitiveness Relation." *The Journal of Economic Perspectives*, Vol. 9, No. 4, pp. 97–118, 1995.

Ridenour, Amy. "Environment: Are Conservatives 'Un-American' on Global Warming?"

Rosenfeld, A. "Sustainable Development, Step 1: Reduce Worldwide Energy Intensity by 2 Percent Per Year." Powerpoint presentation to International conference on Enhanced Building Operations Berkeley, CA 10-14-03

Ross, M., R. Socolow. et al. "Efficient Use of Energy: A Physics Perspective." *American Physical Society*, January 1975.

Samuelson, Paul A. and William D. Nordhaus. *Economics (12th Edition)*. New York: McGraw-Hill, 1985.

Schumacher, E. F. *Small Is Beautiful*. New York: Harper & Row, 1973.

Setty, Gautam and Randall Dodd, "Credit Rating Agencies: Their Impact on Capital Flows to Developing Countries." Financial Policy Forum, Derivatives Study Center, Special Policy Report 6, 2003.

Stavins, Robert and Bradley Whitehead. "Market-Based Environmental Policies," in *Thinking Ecologically*, 1997 [pp. 175–181]

Stiglitz, Joseph. "The Ethical Economist." *Foreign Affairs,* November/

December 2005, pp. 128–134, Vol. 84, No. 6. (The essay is a review of "The Moral Consequence of Economic Growth," by Benjamin Friedman.)

Thaler, R. H. "Toward a Positive Theory of Consumer Choice." *Journal of Economic Behavior and Organization.* 1: 39–60, 1980.

Twain, Mark. *The Gilded Age.*

U.S. Environmental Protection Agency. "National Air Pollutant Emissions Trends 1970–98."

———. "National Emissions Inventory."

U.S. Office of Management and Budget, Office of Information and Regulatory Affairs. Report to Congress On the Costs and Benefits of Federal Regulations, September 1997 and 1998.

Waldman, Peter. "Mercury and tuna: U.S. advice leaves lots of questions." *The Wall Street Journal,* August 01, 2005.

Warren-Alquist Act.

Will, George F. "Our Fake Drilling Debate." *The Washington Post,* December 15, 2005

Wilson, Alex, Jennifer Thorne, and John Morrill. *Consumer Guide to Home Energy Savings.* Washington, DC, American Council for an Energy Efficient Economy, 2003.

Zickel, Raymond E., ed. "Soviet Union" a country study. Federal Research Division, Library of Congress, 1989. See http://lcweb2.loc.gov/frd/cs/sutoc.html

Zingales, Luigi and Raghuram Rajan. "Saving Capitalism from the Capitalists: Unleashing the Power of Financial Markets to Create Wealth and Spread Opportunity."

Index

Them and Us
Cult Thinking and the Terrorist Threat

by Arthur J. Deikman, M.D.
Foreword by Doris Lessing

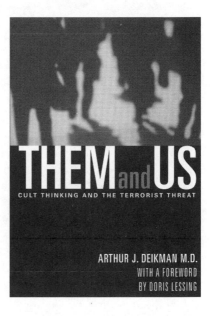

Cult thinking is not something out there—a rare affliction that infects a few people on the margin of society—but a disturbing phenomenon that most of us have experienced in some degree. In *Them and Us*, Arthur Deikman shows the connection between classic cult manipulation and the milder forms of group pressure that can be found in even the most staid organizations—churches and schools, mainstream political movements and corporate boardrooms. In her foreword, Doris Lessing discusses the implications and repercussions of cult thinking on contemporary society.

Arthur J. Deikman, M.D. is a clinical professor of psychiatry at the University of California, San Francisco, and author of *The Observing Self*.

"A highly persuasive, ground-breaking analysis."
—Booklist

"Updated to incorporate discussion of the post-9/11 world, this book assesses the presence and dynamics of cults and cult-think.... Highly recommended for most public and all academic libraries."
—Library Journal

"Deikman persuasively links cult thinking to patterns of behavior and thought found in everyday life and, with no qualitative differences, to the terrorist groups that threaten that life."
—Publishers Weekly

$17.95 trade paperback, 240 pages, 6" X 9" ISBN: 0-9720021-2-X

Get Slightly Famous

Become a Celebrity in Your Field and Attract More Business with Less Effort

by Steven Van Yoder

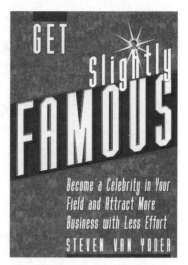

With practical marketing help for the small business owner or independent professional on every page, Get Slightly Famous shows how to tap the secret everyone knows but few practice: It's easier to attract clients through your reputation than sell someone who has never heard of you. Steven Van Yoder describes how any business owner can break out of the sea of competing look-alikes to become "slightly famous."

Steven Van Yoder is a PR practitioner and freelance journalist whose work has appeared in more than 200 publications.

"Get Slightly Famous is more than just slightly valuable, more than just slightly fun to read, and more than just slightly wonderful. It is loaded with insights I wish I had when I was first starting out. But I'm delighted to get them now, and I'll bet every reader will feel the same."
—Jay Conrad Levinson, author of the *"Guerrilla Marketing"* series

"With practical ideas and inspirational success stories, this is a must-read for entrepreneurs. The book's underlying premise and promotional strategies will make readers both more memorable and more money."
—PR Week

"Yoder's approach can be applied to many different disciplines. What's required is careful planning, hard work and patience."
—Miami Herald

$18.95 trade paperback, 280 pages, 6" X 9" ISBN: 0-9720021-1-1
Available at www.BayTreePublish.com

Take Back Your Life
Recovering from Cults and Abusive Relationships

by Janja Lalich and Madeleine Landau Tobias

Cult victims and those who have suffered abusive relationships often suffer from fear, confusion, low-self esteem, and post-traumatic stress. *Take Back Your Life* explains the seductive draw that leads people into such situations, provides insightful information for assessing what happened, and hands-on tools for getting back on track. Written for the victims, their families, and professionals, this book leads readers through the healing process.

Janja Lalich, Ph.D., is Associate Professor of Sociology at California State University, Chico. She is the author of *Bounded Choice: True Believers and Charismatic Cults*, and co-author, with Margaret Singer, of *Cults in Our Midst*.

Madeleine Tobias, M.S., R.N., C.S., is the Clinical Coordinator and a psychotherapist at the Vet Center in White River Junction, Vermont, where she treats veterans who experienced combat and/or sexual trauma while in the military.

"Former members of cults (including those born or raised in such groups) and individuals coming out of abusive relationships will gain valuable insights from this book on how to deal with those experiences and their aftermath. Concerned professionals, family members, and friends will also find useful advice on aiding those who have been abused. I believe that *Take Back Your Life* is the most insightful practical introduction to this subject."
—Michael D. Langone, Ph.D.
Executive Director, International Cultic Studies Association

Take Back Your Life is must reading for everyone who wants to understand the powerfu pappeal that cults have for so many ordinary people, using so many disguises, with so many subtle tactics. This book's wisdom is vital for us all.
—Philip G. Zimbardo, Ph.D., Professor of Psychology, Stanford University

$19.50 paperback, 376 pages, 6" X 9" ISBN: 0-9720021-5-4
Available at www.BayTreePublish.com